Stimulation for Faecal Incontinence

Stimulation for Faecal Incontinence

PhD Thesis by

Jonas Worsøe

Center for Sensory-Motor Interaction (SMI),
Department of Health Science and Technology,
Aalborg University, Aalborg, Denmark

River Publishers

Aalborg

ISBN 978-87-92982-12-4 (paperback)
ISBN 978-87-92329-53-0 (e-book)

Published, sold and distributed by:
River Publishers
P.O. Box 1657
Algade 42
9000 Aalborg
Denmark

Tel.: +45369953197
www.riverpublishers.com

Table of Contents

Preface

This dissertation is based on the work carried out between September 2007 and August 2010 at the Center for Sensory-Motor Interaction, Aalborg University, Denmark. Experiments were performed at the Surgical Research Unit, Department of Colorectal Surgery, Aarhus University Hospital, Aarhus, Denmark.

This work was made possible by financial support from The Danish National Advanced Technology Foundation, Copenhagen, Denmark.

A number of people should be acknowledged for contributing to the completion of this work.

Primarily I thank the patients who participated in the experiments, and without whom nothing of this would have been possible. I would like to thank my main supervisor Nico Rijkhoff for providing me thorough supervision and guiding my scientific education. I also want to thank my co-supervisor Søren Laurberg, Department of Surgery P, Aarhus University Hospital for a relentless enthusiasm, support and constructive criticism. Special thanks go to my other co-supervisor, Klaus Krogh, Department of Hepatology and Gastroenterology, Aarhus University Hospital for providing outstanding guidance, a superb ability to put things in perspective, and for always being available when things were challenging.

Big thanks to colleagues and friends at the Surgical Research Unit, Department of Surgery P for excellent backing, and for creating a helpful and friendly environment. I also wish to give my warmest thanks to the staff at the Anal Physiology Laboratory, Aarhus University Hospital for always being positive and for assisting with the experiments in the most qualified manner.

A special thanks to my parents, Ulla and Peter Henrik, and to my brother Kasper. Finally, I wish to thank my dear girlfriend Sofie for love and support, and for encouraging me when that was needed.

Jonas Worsøe
Aarhus, September 2011

Abstract

The prevalence of fecal incontinence is approximately 5% in the general population. Fecal continence depends on interactions between stool consistency, colorectal motility, rectal capacity and compliance, anorectal sensitivity, pelvic floor muscles and the function of the internal and external sphincter muscles. There are several known risk factors for faecal incontinence (FI), but in many patients no etiology can be recognized and the condition is termed idiopathic fecal incontinence. Patients with spinal cord injury often have neurogenic bowel problems with varying symptoms depending on the severity and level of the lesion. In all patients, the treatment of fecal incontinence is primarily conservative and surgery is offered as second line treatment. Electrical stimulation of the sacral nerves (SNS) with an implanted electrode is a minimally invasive and effective treatment of idiopathic FI and the indications for SNS are widening. However, SNS requires surgery while percutaneous electrical stimulation of peripheral nerves could be a non-invasive alternative.

The dissertation comprises five studies. Study I examined the clinical effectiveness of transcutaneous stimulation over 3 weeks. Study II and III examined the acute effect of rectal stimulation in idiopathic incontinent patients and patients with spinal cord injury. In study IVthe validity of a new research tool for assessment of gastrointestinal motility during stimulation was evaluated. In study V, small intestinal motility was investigated in patients implanted with a sacral nerve stimulator for feacal incontinence.

Study I was a pilot study investigating stimulation of the dorsal genital nerve (DGN) as treatment for idiopathic fecal incontinence. Nine patients completed 3 weeks of treatment with DGN stimulation twice daily. Compared with a 3-week baseline period, the study demonstrated a significantly lower Wexner score (p = 0.027) and St. Mark's score (p = 0.035) after treatment. The number of incontinence episodes obtained from bowel habit diaries was significantly reduced during treatment (p = 0.025). Three weeks after DGN stimulation, the Wexner score (p = 0.048) and the St. Mark's score (p = 0.049) were still significantly lower and the number of incontinence episodes was lower than at baseline (p = 0.017).

In Study II the acute effect on the rectum during DGN stimulation was investigated in patients with idiopathic fecal incontinence. It was hypothesized that DGN stimulation would inhibit rectal tone and increase capacity. Ten patients were included. Pressure-controlled distensions were performed and rectal cross sectional areas were measured during DGN stimulation. Comparisons were made with control distensions without stimulation. No significant differences in rectal cross sectional area, wall tension or compliance were observed.

Study III examined the acute effect on the rectum during DGN stimulation in patients with supraconal spinal cord injury and neurogenic bowel dysfunctions. Again, it was hypothesized that DGN stimulation would inhibit rectal tone and increase capacity. Seven patients were included. In a randomized design rectal cross sectional areas were measured during rectal distension with and without DGN stimulation. In contrast to our hypothesis, rectal cross-sectional areas were significantly lower during DGN stimulation compared to control distensions at distension pressures of 20 cmH2O (p=0.02) and 30 cmH2O (p=0.03). Compliance was also significantly decreased during DGN stimulation at distensions with 20 cmH2O (p = 0.03) and 30 cmH2O (p = 0.04).

In study IV, the Magnet Tracking System (MTS-1) for assessment of gastrointestinal transit and motility patterns was evaluated in eight healthy volunteers. Gastric emptying and small intestinal transit time were determined using MTS-1 and capsule endoscopy. There was good agreement and no systematic difference between the methods when assessing both gastric emptying (median difference 1 min) and small intestinal transit time (median difference 0.5 min). Additionally, it was demonstrated that MTS-1 could separate between fasting and postprandial small intestinal motility. The mean contraction frequency in the small intestine was significantly lower in the fasting state than in the postprandial state (9.90 min-1 vs. 10.53 min-1) (p=0.03). The mean contraction frequency decreased during the first two hours after pyloric passage both during fast (-0.49 min-1) and postprandially (-1.12 min-1) (p=0.04).

In study V MTS-1 was used to investigate idiopathic FI patients implanted with a sacral nerve stimulator. Median gastric emptying was 10 min during SNS and 19 min during noSNS (ns). The mean gastric contraction frequency was 3.06 min-1 during SNS and 3.15 min-1 without SNS (ns). The mean contraction frequency in the small intestine for two hours during SNS was 10.12 min-1 and 10.07 min-1 during noSNS (ns). The mean contraction frequency decreased during the first two hours after pyloric passage both during SNS (-0.98 min-1) and without SNS (-0.79 min-1) (ns).

Based on the present thesis it can be concluded:

1). DGN stimulation reduces fecal incontinence in patients with idiopathic feacal incontinence but the mode of action remains unknown.

2). DGN does not affect rectal wall properties in patients with idiopathic faecal incontinence.

3). DGN increases rectal tone in patients with supraconal spinal cord injury, but the validity and clinical importance of the finding need further study.

4). MTS-1 is useful for determination of gastric emptying and small intestinal transit time.

5). Preliminary results indicate that SNS does not affect small intestinal motility, but data need further analysis and have to be confirmed in a larger study.

Dansk resume

Prævalensen af fækal inkontinens er i baggrundsbefolkningen ca. 5%. Der findes flere kendte risikofaktorer, men hos en del patienter findes ikke nogen årsag. Kontinens afhænger af et samspil mellem afføringens konsistens, colorectal motilitet, rectal kapacitet og compliance, anorectal sensibilitet, bækkenbundsmuskulaturen og funktionen af den interne og eksterne anale sphincter. Patienter med rygmarvsskader har ofte tarmproblemer med vekslende symptomer afhængig af skadesniveau og gruppen er ofte svære at behandle. Behandlingen af fækal inkontinens er primært konservativ og i anden række kan patienter tilbydes kirurgi. Elektrisk stimulation af sakralrødder med en indopereret elektrode er en effektiv behandling. Elektrisk stimulation af perifære nerver kan være et mindre indgribende alternativ.

Afhandlingen omfatter fem studier. Studie I undersøgte den kliniske effekt af transkutan stimulation over 3 uger. Studie II og III undersøgte den akutte effekt på rectum ved stimulation hos idiopatisk inkontinente patienter og hos patienter med rygmarvsskader. Studie IV evaluerede en nyt metode (Magnet Tracking System) til undersøgelse af gastrointestinal transit og tarmmotilitet. I studie V undersøgtes tyndtarmsmotiliteten hos patienter implanteret med en sakral nerve stimulator.

Studie I var et pilot-studie med stimulation af den dorsale genitale nerve (DGN) til behandling af idiopatisk fækal inkontinens. Ni patienter gennemførte 3 ugers behandling med DGN stimulation to gange dagligt. Sammenlignet med en 3 ugers kontrolperiode før behandling var der i behandlingsperioden signifikant forbedring af Wexner (p=0.027) og St. Marks (p=0.035) scores for fækal inkontinens. Endvidere var der en signifikant reduktion i antallet af inkontinensepisoder bedømt med dagbog (p=0.025). Tre uger efter endt DGN stimulation var der stadig signifikant bedre Wexner score (p=0.048) og St. Marks score (p=0.049) samt signifikant færre inkontinensepisoder bedømt med dagbog (p=0.017).

I studie II undersøgtes den akutte effekt på rectum under DGN stimulation hos patienter med idiopatisk fækal inkontinens. Hypotesen var, at DGN stimulation ville hæmme tonus i rectum og øge kapaciteten. Ti patienter blev undersøgt. Under tryk-kontrollerede distensioner måltes rectale tværsnitsarealer med DGN stimulation og disse blev sammenlignet med kontrol distensioner uden

DGN. Der kunne ikke påvises en signifikant ændring af tværsnitsarealer under DGN stimulation sammenlignet med kontroller. Der var heller ingen signifikant ændring af vægtension eller compliance.

I **studie III** undersøgtes den akutte effekt på rectum under DGN stimulation hos patienter med supraconal rygmarvsskade og neurogen tarmdysfunktion. Hypotesen var, at DGN stimulation ville hæmme tonus i rectum og øge kapaciteten. Syv patienter blev undersøgt. I et randomiseret design sammenlignedes rectale tværsnitsarealer under tryk-kontrollerede distensioner henholdsvis med og uden DGN stimulation. Under DGN stimulation var tværsnitsarealet i rectum significant mindre ved distensionstryk på 20 cmH$_2$O (p=0.02) og 30 cmH$_2$O (p=0.03) over basaltrykket end uden stimulation. Nedsat tværsnitsareal under distension er udtryk for øget tonus i rectum. Tilsvarende var compliance også significant mindre ved DGN stimulation 20 cmH$_2$O (p=0.03) og 30 cmH$_2$O (p=0.02) over basaltrykket, mens der ikke var nogen ændring i vægtensionen.

I **Studie IV** evaluerede vi et nyt system (Magnet Tracking System, MTS-1), hvorved man følger en magnet gennem mave-tarm-kanalen. Ventrikeltømning, tyndtarmspassage og motilitetsmønstre i tyndtarmen blev evalueret hos otte frivillige dels i faste og dels efter et måltid. Ventrikeltømning og tyndtarmspassage blev bestemt med MTS-1 og kapselendoskopi. Der var god overensstemmelse og ingen systematisk forskel mellem de to metoder for både ventrikeltømning (median forskel 1 min, range: 0-6 min) og tyndtarmspassage (median forskel 0.5 min range: 0-52 min). Med MTS-1 var det muligt at skelne mellem motilitet i tyndtarmen under faste og postprandielt. Den gennemsnitlige kontraktionsfrekvens var signifikant lavere under faste sammenlignet med efter et måltid (9.90 min^{-1} vs. 10.53 min^{-1}) (p=0.03). Den gennemsnitlige kontraktionsfrekves faldt i løbet af 2 timer både under faste (-0.49 min^{-1}) og postprandielt (-1.12 min^{-1}) (p=0.04).

I **studie V** anvendtes MTS-1 til undersøgelse af tyndtarmsmotilitet hos idiopatisk inkontinente behandlet med sakral nerve stimulation (SNS). Den mediane ventrikeltømning var 10 min med og 19 min uden stimulation (ns). Den median kontraktionsfrekvens i ventriklen var 3.06 min^{-1} med og 3.15 min^{-1} uden stimulation (ns). Den gennemsnitlige kontraktionsfrekvens i tyndtarmen var også uændret med (10.005 min^{-1}) og uden (10.09 min^{-1}) stimulation (ns). Den gennemsnitlige kontraktionsfrekves i tyndtarmen faldt i løbet af 2 timer både under stimulation (-0.965 min^{-1}) og uden stimulation (-0.845 min^{-1}) (p=0.04).

I denne afhandling blev det vist, at DGN stimulation reducerer inkontinens symptomer hos idiopatisk inkontinente og dette bør undersøges nærmere i et randomiseret studie. Patofysiologiske studier under akut stimulation hos idiopatisk inkontinente patienter var ikke i stand til at demonstrere en effekt på rectum og en evt. virkningsmekanisme er fortsat uafklaret. Patofysiologiske studier under akut stimulation hos patienter med rygmarvsskade viste en øget tonus i rectum. Betydningen af dette fund og om det på sigt kan gavne rygmarvskadede patienter med tarmdysfunktion er ikke afklaret. Endelig blev MTS-1 anvendt som et nyt redskab til bestemmelse af transit tider og motilitet i ventrikel og tyndtarm. Fordelene ved MTS-1 er, at metoden er non-invasiv og uden strålingsbelastning.

Til sidst blev MTS-1 anvendt til undersøgelse af tyndtarmsmotilitet hos idiopatisk inkontinente patienter implanteret med SNS, hvor der fandtes ikke nogen påvirkning af tyndtarmen ved stimulation.

Abbreviations

ARP	Anal resting pressure
ASP	Anal squeeze pressure
ASR	Anal sampling reflex
CSA	Cross sectional area
DGN	Dorsal genital nerve
EAS	External anal sphincter
FI	Fecal incontinence
HAPC	High amplitude propagated contraction
IAS	Internal anal sphincter
IP	Impedance planimetry
NBD	Neurogenic bowel dysfunction
PRMA	Periodic rectal motor activity
PTNML	Pudendal terminal nerve motor latency
PTNS	Posterior tibial nerve stimulation
RAIR	Recto-anal inhibitory reflex
SCI	Spinal cord injury
SNS	Sacral nerve stimulation

Chapter 1

Introduction

1.1 BACKGROUND

Fecal incontinence (FI) is the loss of regular control of bowel movements and the inability to defer defecation to a proper time and a proper place. A formal definition of functional FI is provided by the Rome II criteria as the "recurrent uncontrolled passage of fecal material for at least one month in an individual with a development age of at least 4 years" [1]. The symptoms are often socially unacceptable, and those affected may be beset by feelings of shame and humiliation. If the problems are not addressed, this can lead to social withdrawal and isolation.

Management of FI is mainly conservative, and in more severe cases surgical procedures represent the second line treatment. Therapeutic deployment of electrical stimulation of the nervous system could represent an intermediary minor invasive or even non-invasive treatment if conservative treatment fails. Electrical stimulation can be applied in various ways. With sacral nerve stimulation (SNS), a stimulator and an electrode are implanted for stimulation of the sacral roots. Stimulation can also be applied peripherally accessing the nerves with a needle electrode (percutaneous) or with self-adhesive plaster electrodes (transcutaneous). In 1981 Tanagho and Schmidt introduced sacral nerve stimulation for the treatment of lower urinary tract dysfunction including urinary urge incontinence [2-5]. Matzel et al. adapted SNS for idiopathic FI in 1995 [6]. Now there is good evidence that SNS benefits patients with FI and new indications for SNS are emerging. These include chronic constipation, irritable bowel syndrome and abdominal pain. An effect from SNS on such diverse conditions can not entirely be explained from stimulation of efferent fibers from the sacral spinal cord. Also, there are strong indications that SNS affects motility in the right side of the colon. This cannot be explained from direct stimulation of visceral motor fibers only. The main effects of SNS are caused not by stimulation of somatic or visceral efferent fibers but rather by neuromodulation through stimulation of either somatic or visceral afferent fibers. Furthermore, it remains to be clarified if SNS has an effect on the upper gastrointestinal tract, i.e. the stomach and the small intestine.

Transcutaneous stimulation of the dorsal genital nerve (DGN), an afferent branch of the pudendal nerve, allows stimulation of somatic afferent fibers. DGN stimulation has an effect on bladder function and preliminary results indicate that it may have an effect on bowel function too. However, the feasibility of DGN stimulation in the treatment of FI and the acute effects on bowel function need to be investigated.

1.2 AIMS

This project was part of the project "Implantable Neural Prostheses" initiated by Neurodan and Center for Sensory-Motor Interaction at the Department of Health Science and Technology at Aalborg University, and supported by the Danish Advanced Technology Foundation. The overall aim was to develop nerve prostheses to help patients suffering from urinary incontinence and FI. In this context, the feasibility and effects of DGN stimulation in the treatment of FI were investigated in patients suffering from idiopathic FI and in patients with SCI. Primarily, the effect of therapeutic stimulation of the DGN on FI symptoms was evaluated in idiopathic fecal incontinent patients (study I). Secondly, the mechanism of action was explored. The acute effect of DGN stimulation was examined in patients with idiopathic FI (study II) and in patients with a complete supraconal SCI (study III). Next, the validity of a new research tool (Magnet Tracking System, MTS-1) for determination of oroceacal transit times and small intestinal motility was investigated. Finally, preliminary data on gastric and small intestinal motility were obtained using MTS-1 in patients implanted with a sacral nerve stimulator.

Hypotheses:

1. Transcutaneous electrical stimulation of the DGN applied twice daily for three weeks reduces symptoms in patients with idiopathic FI.

2. Transcutaneous electrical stimulation of the DGN applied twice daily for three weeks improves anal resting pressure and squeeze pressure, and increases rectal volume capacity.

3. Acute transcutaneous electrical stimulation of the DGN increases rectal capacity in patients with idiopathic FI.

4. Acute transcutaneous electrical stimulation of the DGN increases rectal cross sectional area in patients with fecal incontinence caused by supraconal SCI.

5. The Magnet Tracking System (MTS-1) allows accurate and reproducible determination of gastric emptying and small intestinal transit time.

6. Differences in fasting and postprandial small intestinal motility can be determined using MTS-1.

7. SNS changes gastric emptying, small intestinal transit and contraction patterns.

The specific aims of each study were:

Study 1: To investigate the tolerability and the efficacy of chronic transcutaneous electrical stimulation of the DGN in patients with idiopathic FI.

Study 2: To investigate the acute effect of transcutaneous electrical stimulation of the DGN, on rectal wall properties in patients with idiopathic FI.

Study 3: To investigate the acute effect of transcutaneous electrical stimulation of the DGN, on rectal wall properties in patients with fecal incontinence caused by complete supraconal SCI.

Study 4: To determine if MTS-1 can provide valid information on gastric emptying and small intestinal transit time in healthy subjects. Further, to study if MTS-1 can differentiate between fasting and postprandial small intestinal motility in healthy subjects.

Study 5: To present preliminary data on gastric emptying, small intestinal transit and small intestinal motility patterns in patients with FI implanted with a sacral nerve stimulator.

1.3 ANATOMY

The present thesis includes studies on gastric, small intestinal, rectoanal and pelvic floor function. In the following the anatomy and physiology of these structures are briefly described.

1.3.1   Stomach

The stomach serves as a reservoir (volume: 500-1000 ml) connecting the oesophagus and the small intestine. It can be sealed off by the lower oesophageal sphincter proximally and the pyloric sphinter distally. There are four anatomical regions of the stomach (the fundus, corpus, antrum and pylorus), and the wall has three layers of smooth muscle cells (outer longitudinal, circular layer, and inner oblique layer). The epithelium of the stomach contains 15-20 millions glands responsible for secretion of enzymes, acid, and electrolytes (1-3 l day-1) [7,8].

1.3.2   Small intestine

The small intestine is situated between the pyloric sphincter and the ileoceacal junction. It is divided into the duodenum, the jejunum, and the ileum and the total length is approximately 3-4 m. With muscular tone the diameter of the small intestine is 2-3 cm. The wall has two layers of smooth muscle cells (outer longitudinal and inner circular layer). The main purpose of the small intestine is digestion and absorption of chyme arriving from the stomach. The epithelium is optimized for this purpose with villi and microvilli increasing the absorptive area from 0.5 m2 to 50-100 m2. Furthermore secretion by the liver, the pancreas and small intestinal glands aggregate to about 3 l day-1. The small intestine also serves an immunological function as 70% of the body's immune cells are in the small intestine [7,8].

1.3.3   Rectum

The rectum is a continuation of the sigmoid, and starts by definition where the mesosigmoid ends. It is 12 to 15 cm long (15 cm by definition used by surgeons), with a diameter between 2-3 cm and 7-8 cm depending on tone. It has three lateral curves, corresponding intraluminally to the valves of Houston. Rectal mucosa is smooth and transparent with columnar epithelium. The inner circular muscle layer, at its lower margin, gives rise to the internal anal sphincter (IAS). The outer longitudinal muscle layer is a continuous layer. Contrary to the colon, rectum lacks appendices epiploicae, haustra and taeniae. Usually the rectum is extraperitoneal on its posterior aspect [9,10]. It is closely related to the uterine cervix and posterior vaginal wall in women, and the bladder and vas deferens in men. At the level of the pelvic floor it continues in the anal canal.

1.3.4   The anal canal

The anal canal is a complex structure including visceral and somatic components. The functional anal canal is approximately 4 cm long, extending from the anal verge to the ano-rectal ring (the upper surface of the pelvic floor), where it meets the rectum [11]. The dentate line is a serrated line with anal valves and sinuses, and the bases of the anal columns. It represents the junction between the different origins of venous and lymphatic drainage and nerve supply. Most proximal, the mucosa consists of single layer columnar epithelium, similar to rectal mucosa. Right above the dentate line is a transition zone (5 to 10 mm) with multi layered cuboidal cells, and below the dentate line, there is modified squamous epithelium. The submucous layer includes hemorrhoidal plexuses and nerves. The IAS is a 25 to 40 mm long (longer in males) condensation of the most distal part of the circular smooth muscle of the rectum. Proximally, it is 2-3 mm thick, and thickens distally to about 5 mm [9]. The outer longitudinal muscle mixes with fibers of the levator ani at the level of the anorectal ring to form the conjoined longitudinal muscle,

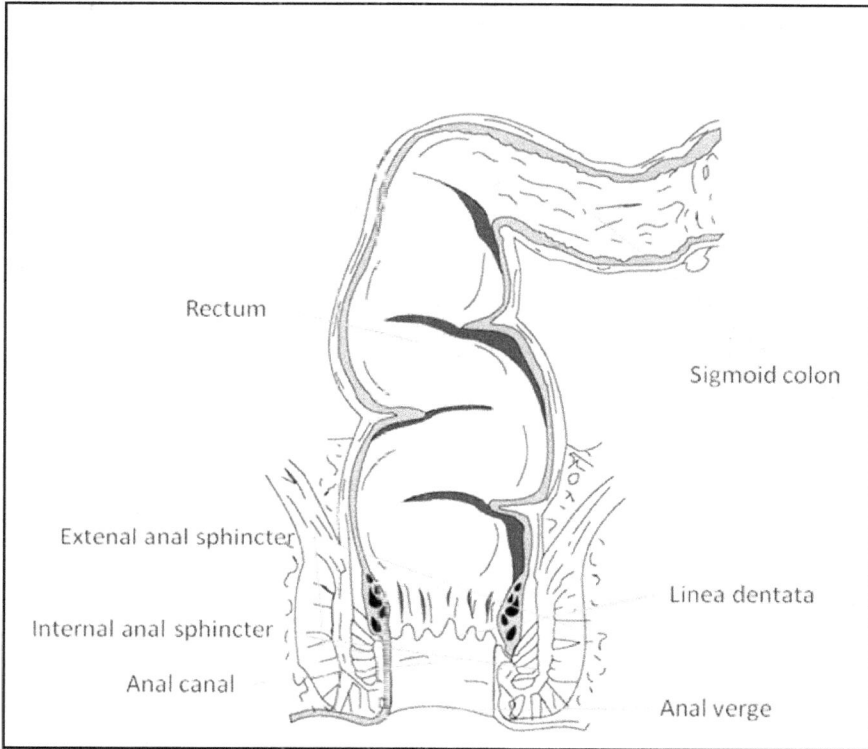

Figure 1.1: Rectum and the anal canal

between the internal and the external sphincter, and attaching the rectum to the pelvis. The external anal sphincter (EAS) is a cylinder of striated muscle which envelopes the entire internal sphincter but ends more distal [9]. At the lower margin of the IAS, a rounded edge marks the boarder to EAS - the intersphincteric line. The EAS includes three parts; the deep part which is in close relation with the puborectalis muscle, the superficial part which is shaped like an ellipse and is attached to the anococcygeal ligament posteriorly and the centrum tendineum perinea anteriorly, and the subcutaneous part which is ring shaped and situated just above the anus.

1.3.5   The pelvic floor

The pelvic floor is mainly composed of the levator ani muscle. It is a muscle sheet including three striated muscles: ileococcygeus, pubococcygeus and puborectalis. In the midline, rectum, urethra and the dorsal penile vein or the vagina pass through the levator hiatus. The levator ani is supplied by the S2-S4 sacral roots. The pelvic floor separates the pelvic cavity above from the perineal region (including perineum) below, and supports the pelvic organs.

1.3.6   Control of gastrointestinal motility

1.3.6.1   Intrinsic innervation

The enteric nervous system (ENS) contains 80-100 million neurons organized in two ganglionated plexes, the myenteric plexus (Auerbach) situated between the outer longitudinal and inner circular muscle layers and the submucous (Meissner) plexus situated between the circular muscle layer and the muscularis mucosa [12]. The ENS, or "the intestinal minibrain", controls motility, with the myenteric plexus mainly regulating smooth muscle activity and the submucosal plexus regulating secretion. The ENS can function independently from the central nervous system but it also integrates local information from the gastrointestinal tract with central commands through postganglionic sympathetic and parasympathetic fibers [13]. A number of neurotransmitters are used by the enteric nervous system. Acetylcholine is widely utilized, but also nitric oxide, substance P, vasoactive intestinal polypeptide, adenosine-triphosphate , neuropeptide Y, dopamine, serotonin, and γ-aminobutyric acid have been demonstrated [12].

1.3.6.2   Extrinsic innervation

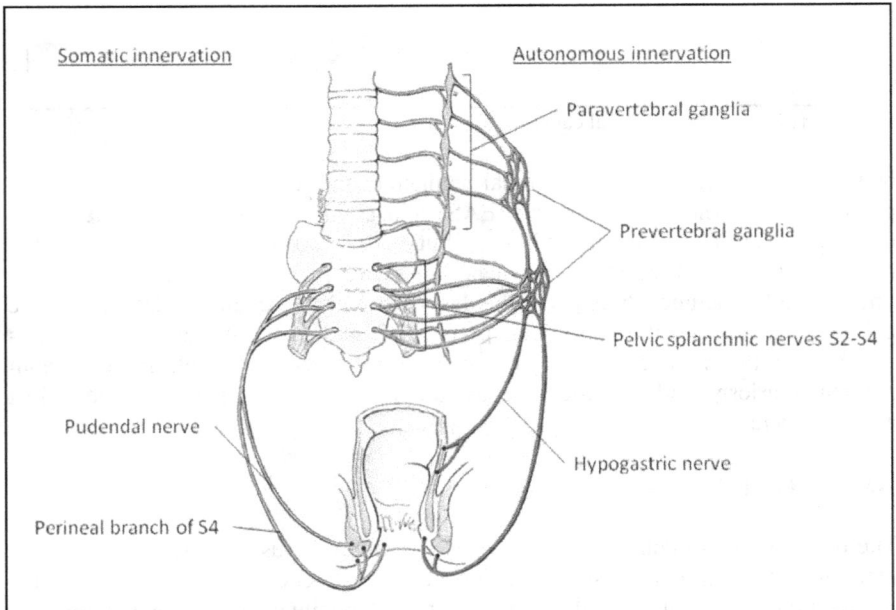

Figure 1.2: Innervation of rectum and the anal canal.

The gastrointestinal tract is innervated dually by the autonomous nervous system. The parasympathetic part promotes peristaltic activity, vasodilatation and secretion while the sympathetic part reduces peristaltic activity and vasodilatation. Sphincters, including the IAS, react oppositely, contracting with sympathetic stimulation and relaxing with parasympathetic stimulation. The stomach and small intestine receive parasympathetic innervation from the vagal nerve. Sympathetic innervation of the stomach comes from the splanchnic thoracic nerves (preganglionic fibers) connecting to postganglionic fibers in the plexus coeliacus. Postganglionic sympathetic fibers for the small intestine are from the superior mesenteric plexus. Parasympathetic innervation of the distal part of the colon (below the spleenic flexure) and the anorectum are provided by the sacral roots (S2-S4) (Fig. 2.1). Preganglionic splanchnic pelvic nerves run to the intestinal wall through the inferior hypogastric plexus on the lateral aspect of the rectum. From the pelvic plexus, the hypogastric nerves innervate the IAS, and ascend to innervate the colon. Sympathetic innervation of the ano-rectum arises from the ninth thoracic to the third lumbar segments of the spinal cord. Preganglionic fibers synapse in the paravertebral ganglias (the sympathetic chain) or arrive at the mesenteric ganglia (prevertebral ganglia) passing through the sympathetic chains. Postganglionic fibers continue as sacral splanchnic nerves along small vessels to the pelvic plexus and innervate the intestinal tract. The neurotransmitter in preganglionic sympathetic neurons and in both preganglionic and postganglionic parasympathetic neurons is acetylcholine. The neurotransmitter in postganglionic sympathetic neurons is noradrenalin. Cell bodies of spinal afferents are located in dorsal root ganglia and project to the dorsal horn of the spinal cord and the dorsal column nuclei. Spinal afferents are broadly subdivided into splanchnic and pelvic afferents that follow the paths of sympathetic and parasympathetic efferents to the gut wall. Somatic afferents, which innervate the striated musculature of the pelvic floor, project to the sacral spinal cord via the pudendal nerve. The IAS is innervated by sympathetic (L5) and parasympathetic (S2-S4) nerves. The pelvic floor and voluntarily controlled EAS is innervated on both sides by somatic fibers in the pudendal nerve (S2-S3) and the perineal branch of the S4 sacral nerve (Fig. 1.1 and Fig. 2.1) [14].

1.3.6.3   Humoral control

Colorectal motor function is influenced by luminal content, immunological factors and hormones. Thyroid hormones promote colorectal motility. Epinephrine reduces colorectal motility. Serotonin is expressed by enterochromaffin cells, and is the major paracrine mediator [15]. Enteric neurons and sensory afferents have 5-HT receptors, which mediate excitatory actions by serotonin including promotion of motility, secretion and sensation [16]. The immune system is represented by several types of cells including leucocytes, lymphocytes, dendrocytes, macrophages, and mast cells. They are situated in the mucosa and smooth muscle in proximity to the intrinsic neural network and extrinsic fibers, which allow them to influence and modulate neural function. Mast cells can recognize foreign agents (by sensitized immunoglobulin E antibodies attached to the surface) and release various mediators including histamine. Histamine can induce long-lasting excitation in the enteric neurons promoting motility, increase intestinal secretion by blocking sympathetic innervation, and increase gastrointestinal sensory input to the central nervous system by modulation of splanchnic afferents [17-19].

1.4 NORMAL GASTROINTESTINAL MOTILITY

1.4.1   Physiology of gastrointestinal motility

Gastrointestinal motility is controlled by slow wave activity generated by intestinal pacemaker cells (interstitial cells of Cajal). The cells of Cajal surround the circular muscle layer in a network along the entire gastrointestinal tract [20]. The slow wave frequency differs in different parts of the gastrointestinal tract. In the stomach the slow wave frequency is 3 min-1. The frequency of slow waves is approximately 11 min-1 in the proximal small bowel and 7-8 min-1 in the ileum [21,22]. In the colon and rectum, the slow waves are seen both at a 2-4 min-1 and at a 6-12 min-1 frequency [23-25]. Slow waves can trigger action potentials in the smooth muscle membrane, which in turn result in muscular contraction. Inhibitory neurons control if the slow wave generates an action potential and subsequent muscular contraction occurs. The circular muscles are electrically connected through gap junctions, which allow action potentials to travel between muscle fibers. This enables muscular contractions to appear uniformly around the circumference during peristalsis [8].

1.4.2   Motility of the stomach and the small intestine

The proximal region including the fundus and one-third of the corpus is adapted for storage and retention. The muscles maintain a continuous tone and can relax in response to filling of the stomach (stress relaxation). The distal region (two-third of the corpus, pylorus) is specialized for mixing and grinding facilitated by phasic contractions. The contractions occur in rhythmic cycles at a frequency of 3 min-1. As propulsive contractions originating in the proximal distal region propel the content towards the closed pyloric channel the pressure increases and retropulsion sends the contents back into the proximal stomach. This contributes to digestion and grinding of food particles.

Small intestinal fasting motility comprise phase I, II and III of the migrating motor complex [26]. Phase I is a period of silence present for about sixty percent of a cycle length. The propagating contractions are initiated in the distal region of the stomach. Phase II appear for about thirty percent of the cycle and is a period of irregular contractions. Phase III is the hallmark of fasting motility. The purpose is to intermittently clear the small intestine of secretions, debris, and bacteria. Phase III migrates at a velocity of 5 to 10 cm min-1 [27,28]. Phase III usually lasts 2 to 3 minutes and seldom more than 10 minutes. It returns at average intervals of 85-110 minutes [27,28]. Even so, apart from progression velocity, large variability characterize most MCC variables [29,30]. Postprandial motility starts 5 to 10 minutes after ingestion and continues as long as food remains in the stomach [31]. The length of the postprandial period depends on the caloric content of the meal usually lasting for 2-3 hours. In the feed state contractions are of variable amplitude and about 50 percent of contractions do not propagate [32].

1.4.3   Ano-Rectal motor activity

Rectal motor activity includes various contraction patterns. There is some coordination with colonic motility. Colonic mass contractions, also termed high amplitude propagated contractions (HAPC), originate orally in the colon, and can propagate all the way to the rectum [33]. Almost all HAPC activity takes place during day time. It is associated with postprandial periods (gastro-colic response) or awakening. The function of HAPC is to move colonic contents forward and when reaching the rectum, they may stimulate a desire to defecate [33]. Periodic rectal motor activity (PRMA) is trains of contraction with a frequency of 2-3 min-1, lasting at least 3 min, and with diurnal variation in recurrence frequency. [34-36]. significantly more PRMAs propagate retrograde, but 34-44% are restricted to the rectum [34]. Data on diurnal variation of PRMAs and relation with ingestion of food have been conflicting [34,35,37,38]. PRMAs are not related to anal sphincter activity, but most follow shortly after colonic motor events [34]. This temporal association suggests that PRMA are triggered by content in the rectum. A predominant retrograde directed nocturnal activity could indicate that PRMA acts as a break or gatekeeper to secure continence [34].

In between defecations, the IAS is in a state of tonic contraction with some phasic variation and creates a natural barrier which accounts for about 30-55 % of the resting anal pressure [9,39,40]. Retrograde contraction waves may contribute to movement of content and prevent soiling [41,42]. Furthermore, there is inconsistent motor activity with spontaneous contractions and relaxations [43]. The rectoanal inhibitory reflex (RAIR) and the anal sampling reflex (ASR) represent the relaxation of the IAS in response to distension and contraction of the rectum respectively [44]. They may allow sampling of rectal content by sensory receptors in the proximal anal canal to discriminate between solid and liquid stool and flatus [45]. The reflexes depend on intrinsic non-adrenergic non-cholinergic neural pathways. An ano-colonic reflex coordination has also been proposed. Relaxation of the IAS was found to precede initiation of HAPCs [46].

1.5 FECAL INCONTINENCE

1.5.1   Factors important for fecal continence

Several factors contribute to the maintenance of fecal continence, and disturbances of one or more of these may result in FI [47]. Stool consistency must preferably be formed rather than liquid since formed stools are much easier controlled than liquid stools. Colorectal motility controls the load of feces into the rectosigmoid. Retrograde propulsion in the rectosigmoid secures that the rectum is empty between defecations. Rectal capacity and rectal compliance is important for the ability of the rectum to accommodate to the fecal load. Anorectal sensibility is important for the perception of fecal content and subsequent activation of the external sphincter and the ability to defer defecation. The IAS is constantly contracted and, together with hemorrhoidal cushions, forms a sealed barrier. Contraction of the pelvic floor muscles also contributes by form a barrier. The puborectalis sling controls the anorectal angle which may also plays a role by creating a flap valve.

1.5.2   Epidemiology of FI

In an American study of FI including over 2308 participants, the prevalence of FI was 8.3%, with 6.2% being incontinent to liquids, 1.6% incontinent to solid stool and 3.1% leaking mucus. There were no gender difference (women 8.7%, men 7.7%), but the prevalence increased with age from 2.6% in 20 to 29 year olds to 15.3% in subjects 70 years and older. No socio-economic risk factors were identified (ethinicty, eductation, income, marital status) [48].The onset rate in community dwelling subjects aged 50 years or older has been found to be 7% per 10 years [49]. Several studies have shown that FI significantly reduces daily function and quality of life [50-52].

1.5.3   Etiology of FI

There are several etiological factors for FI (Table 1). A limited proportion is caused by congenital malformations including imperforate anus, rectal agenesis, meningocele, myelomeningocele, and cloacal defects [53]. The majority of cases result from sphincter disruption during vaginal delivery. FI rates of 9.6% to 17.0% have been reported in postpartum women [54,55]. Obstetric complications associated with increased risks of FI [56]. Sphincter defects not recognized in the immediate postpartum period have been shown in up to 35% of primiparous women after vaginal delivery, and this may cause FI [57-59]. Traction injury to the pudendal nerve, with compromised function of the EAS and the pelvic floor, can also be caused by vaginal delivery [60]. Various anorectal surgical procedures are also etiological factors [61-64]. Colorectal resection may also cause incontinence because of loss of reservoir function and decreased colonic transit time [65].

Congenital	Spina bifida
	Rectal agenesis
	Cloacal defects
	Imperforate anus
Anatomical	Obstetric injury
	Anorectal surgery
	Anorectal and pelvic trauma
	Rectal prolapse
Neurological	Diabetes mellitus
	Multiple sclerosis
	Stroke
	Dementia
	Central nervous system tumor
	Spinal cord injury
	Pudendal neuropathy
Functional	Inflammatory bowel disease
	Irritable bowel syndrome
	Radiation proctitis
	Malabsorption
	Hypersecretory tumors
	Faecal impaction
	Psychiatric disorder
	Physical disability

Table 1.1: Causes of FI (modified from Madoff et al. [53]).

Gastrointestinal dysfunction with increased risk of FI may arise from inflammatory bowel disease, malabsorption, infectious diarrhea, secretory dysfunction, sequelae from irradiation, and irritable bowel syndrome [66-69]. Constipation because of fecal impaction may lead to co-existing overflow incontinence due to reduced anorectal sensibility [70].

Various diseases affect central and peripheral nervous system and may also cause FI. These include diabetes, stroke, dementia, central nervous systems pathology (tumor, trauma, and infection), and multiple sclerosis.

1.5.4   Idiopathic FI

In patients with idiopathic FI no specific etiology is identified. A study among idiopathic fecal incontinent women in the US, found rectal urgency and age to be risk factors, but did not identify specific risk attributable to cesarean section, vaginal delivery or anorectal surgery [71]. In a recent study, the prevalence of idiopathic FI was 2.0% among Canadian community dwelling adults accounting for more than half of the overall prevalence of 3.7% across all genders [72].

1.5.5   Patophysiology of idiopathic FI

No obvious pathophysiological features characterize idiopathic FI. Both the ARP and the ASP overlap with normal values [73]. Among patients with idiopathic fecal incontinence, women have significantly lower anal ARP compared with male patients [74]. The maximum ARP and the anal resting pressure gradient have been found to be significantly lower compared with continent individuals [75]. Threshold for first sensation rectal volume tolerability is also lower [76,77]. Severe straining, in constipated patients, may lead to progressive loss of rectal sensation, which can also result in FI [78]. Pudendal nerve damage may be present and bilateral PTNML may be longer in patients with FI, and may be associated with low ARP, but does not correlate with ASP [79,80].

1.6 NEUROGENIC BOWEL DYSFUNCTION

1.6.1   Bowel symptoms in spinal cord injury patients

Most subjects with SCI experience neurogenic bowel dysfunction (NBD) which usually includes combinations of FI, constipation, impaired defecation, abdominal pain and bloating [81-83]. Approximately 75% of SCI patients suffer from FI, 80% report constipation, 30% regard colorectal problems as worse than both bladder or sexual dysfunction and almost 40% feel restricted on social activities and quality of life [81,82]. The severity of NBD is associated with level of lesion, completeness of lesion, and time since injury (≥ 10 years) [82-85].

The severity of FI is associated with the level of SCI [82,86]. Vallés et al. investigated 54 patients with chronic complete SCI (mean duration since injury six years) and identified three different patterns. Lesions above T7 were associated with delayed segmental colonic transit in the left colon and rectosigmoid, inability to increase intraabdominal pressure and relax the anal sphincters during defecation. This caused significant difficulty with defecation, very frequent constipation, but not very severe FI. Lesions below T7 with preserved sacral reflexes were characterized by severe defecatory difficulties, less severe constipation, and not very severe FI. Lesions below T7 without preserved sacral reflexes were characterized by infrequent constipation, less defecatory problems, and more severe FI [87]. Clinton et al. investigated 110 patients (median age 45 years), with a median time since lesion of 17 years. 74% were constipated, 22% had abdominal bloating, 33% had gastrointestinal pain, and 41% had FI. Abdominal bloating and constipation were associated with cervical and lumbar rather than thoracic level of injury. Abdominal pain and FI were associated with higher anxiety scores [88].

Gastrointestinal dysfunction is associated with duration of injury, even though the development of specific symptoms differs. Krogh et al. found that duration of injury was associated with both FI and autonomous symptoms related to defecation [82]. Lynch et al. found that time spent at toilet and use of laxatives depended on time since lesion, while FI symptoms were unaffected [83]. Stone et al. showed that chronic gastrointestinal problems were infrequent in the first five years, but defecatory problems became more common over time. Abdominal pain and distension increased in long-term injured patients (duration >18 years) [89]. Han et al. reported that bowel habits were "settled" within the first year in 93%, and no correlation was found between duration of injury and bowel dysfunction [85]. In a prospective study, Faaborg et al. described changes in colorectal symptoms during ten years. More patients reported that colorectal dysfunction, had major impact on their quality of life in the second survey (25% vs. 38%. p<0.005). The frequency and severity of constipation related symptoms (frequency of defecation, time spent at defecation and the need for digital anorectal stimulation) were significantly increased, while FI was significantly less frequent [90].

1.6.2   Patophysiology of neurogenic bowel dysfunction

Lesions of the conus medularis or cauda equina directly affect parasympathetic innervation of the left colon (from the spleenic flexure), the rectum, and the IAS. Supraconal lesions above S2 but below L2 spare sympathetic control and the sacral parasympathetic reflex center, but suprasacral inhibition is lost (upper motor neuron lesion). Lesions above L1 affect both supraconal control of parasympathetic innervation and to some degree sympathetic innervation of the left colon, rectum and IAS. Lesions above T6 leave both sympathetic and parasympathetic spinal centers functional, but without supraspinal inhibition [14]. All supraconal lesions result in rectal hypertonia and hypercontractility due to unopposed spinal reflexes [91].

ARP and ASP are reduced in SCI patients with no significant relation to level of injury [92]. The rectal compliance is reduced in patients with supraconal lesions, and increased in patients with conal/ cauda equine lesions [91,93]. Likewise, the response of the IAS to rectal distension is attenuated in conal or cauda equine lesions [91,94]. Colonic transit time is significantly prolonged with supraconal lesions affecting total gastrointestinal transit time and segmental transit throughout the colon, while conal/ cauda equina lesions mainly affect total gastrointestinal transit time and segmental transit time in the descending colon and rectosigmoid [95-97]. In conal/ cauda equine lesions, defecation is impaired because of damage to the motor innervation of the colorectum. Such damage reduces emptying of the rectosigmoid and descending colon during defecation [98]. With supraconal lesions, rectal hyperreactivity and strong rectal contractions, may cause reflex defecation [44]. The hyperreactive rectoanal reflexes can be utilized to trigger defecation by digital stimulation. Accordingly, reflex mediated defecation is more difficult in subjects with conal or cauda equine lesions.

1.6.3   Autonomic dysreflexia

Patients with a complete lesion above T6 can experience autonomic dysreflexia with increased blood pressure, flushing, decreased heart rate and symptoms like chills, sweating, nausea headache and anxiety [99]. Distension of the bladder or bowel, results in afferent inflow to the dorsal horn of the lumbosacral spinal cord which induces a massive sympathetic reflex discharge below the level of the injury. This results in vasoconstriction of the muscular, splanchnic, and cutaneous vascular beds. The resultant paroxysmal hypertension produces a baroreceptor mediated reflex bradycardia accompanied by withdrawal of sympathetic activity above the lesion level with resultant vasodilatation. The condition is potentially lethal and must be considered whenever introducing new treatment modalities for patients with high SCI.

1.7 SOME IMPORTANT METHODS FOR ASSESSMENT OF GASTRIC EMPTYING AND SMALL INTESTINAL MOTILITY

1.7.1   Scintigraphy

The gold standard for determination of gastric emptying and small intestinal transit time is scintigraphy of a solid state meal [100-102]. Gastric emptying is assessed as the percentage retained after two and four hours, and the half emptying time. Retention of more than 10% after four hours is indicative of delayed emptying [103]. Dual labeling of both solid and liquids with each their isotope is commonly used. However, emptying of liquid may be normal until dysmotility reaches an advanced stage. The reproducibility of scintigraphic assessment of gastric emptying is good [104]. Sensitivity and specificity for detecting delayed gastric emptying are high and existing normal values are based on large control groups. The oroceacal transit is obtained with oral intake of isotopes while intravenous administration with excretion in the bile (99mTc-HIDA) can provide a small intestinal transit time. The normal value for small intestinal transit time is between 131 and 322 min (mean: 220.9 ±49 min) [102]. The intraindividual reproducibility is poor (correlation: 0.43) reflecting physiologic variation[102]. The disadvantages of scintigraphy are costs, a cumbersome procedure, and exposure to radiation.

1.7.2   Manometry

Antroduodenal manometry provides information on both the pressure and the patterns of contractions. Tonic contractions are difficult to measure with manometry [32,105]. With 24 hours ambulatory recording, the diurnal variation of the MMC and postprandial responses can be studied in detail [32]. Abnormal of motor patterns can be recognized and are associated with pathology [31]. Small intestinal manometry can distinguish between myopathic and neuropathic motor dysfunction. Myopathic disorders generate normal contractile activity with lower pressure [106]. Neuropathic disorders result in uncoordinated contractions with normal or higher pressure [107,108]. Currently, manometry is the gold standard for determination of small intestinal motility. However, the method is invasive and may disrupt normal motility.

1.7.3   Wireless motility capsule

A wireless motility capsule (WMC), known as the SmartPill, is an ingestible device, which measures three parameters; pH, pressure and temperature. It provides information about gastric emptying, small intestinal transit, and colonic transit (pH, temperature) [109,110]. The SmartPill also provides information about contractions patterns (pressure). In a study by Rao et al., the WMC was able to detect gastrointestinal dysmotility, and Kloetzer et al. were able to separate gastroparetic patients from controls based on the antroduodenal pressure profiles [111,112]. The method is safe, non-invasive and ambulant, eventhough it should not be used when mechanical obstruction is suspected.

1.8   METHODS FOR EVALUATION OF FI

1.8.1   Assessment of FI

A careful assessment of FI should include an assessment of both the severity of incontinence and the impact on daily activities and quality of life [113]. A wide selection of summary scales has been constructed to assess severity, even though none has been validated [114-120]. Impact measures evaluate the effect on physical, psychological, social, and occupational function. Generic questionnaires can be applied in various populations, including both the ill and the well, facilitating comparison between groups [121,122]. Most often, a disease specific impact measure is needed to obtain better sensitivity.

1.8.2   Bowel habit diary

Day-to-day description of bowel habits can be recorded with a three week bowel habit diary [123]. It records the frequency of defecation, the number of days with urge, frequency of urgency, frequency of urge incontinence, frequency of passive incontinence, the number of days with soiling, and the number of days where a pad is used.

1.8.3   Wexner score

The Wexner continence grading scale assesses the severity of FI [120]. It includes five items (incontinence to solid stool, liquids, and air, the need to wear a pad and alteration of lifestyle). Each item is graded between zero and four to give a summary score between zero and twenty. Zero indicates complete continence and twenty indicates complete incontinence. The Wexner score has been demonstrated to correlate well with an expert's clinical impression and can be used to detect changes in response to treatment [119].

1.8.4   St. Marks score

The St. Marks incontinence score is a modification of the Wexner score [119]. It also includes three items about type and frequency of incontinence and one item assessing alteration in lifestyle. Each of these items is graded from zero to four. Three additional items are included: the need to wear a pad or plug (0 or 2), the use of constipating medicine (0 or 2), and the lack of ability to defer defecation for 15 minutes (0 or 4). The total score ranges between zero (complete continence) and 24 (complete incontinence). Not surprisingly, a high correlation has been demonstrated between the St. Mark's score and Wexner score [119,124]. Correlation between the St. Mark's score and patient's subjective perception of bowel control has been found moderately good, across age and gender or the type of incontinence (passive and/ or urge) [125].

1.8.5   The Fecal Incontinence Quality of Life Scale (FIQL, Rockwood score)

The Fecal Incontinence Quality of Life Scale evaluates the quality of life specifically related to FI. It is composed of a total of 29 items; these items make up four scales: Lifestyle (10 items), Coping/Behavior (9 items), Depression/Self-Perception (7 items), and Embarrassment (3 items). The scales range from one to five, with one indicating a lower functional status. FIQL has previously undergone thorough psychometric evaluation [126].

1.8.6   Neurogenic bowel dysfunction score

A disease specific impact score has been constructed to assess severity of NBD [82,127]. Ten items are included and weighted with points to reflect the impact: frequency of bowel movements (0-6), autonomic symptoms (0-2), the use of laxatives (tablets or drops) (each 0-2), time used per defaecation (0-7), frequency of digital stimulation (0-6), frequency of faecal incontinence (0-13), medication against faecal incontinence (0-4), flatus incontinence (0-2), and perianal skin problems (0-3). The total score is between zero and 47, with 47 representing most severe symptoms.

1.9 ROUTINE METHODS FOR ASSESSMENT OF ANO-RECTAL FUNCTION

1.9.1   Endoanal ultrasound

Endoanal ultrasound is an essential part of the initial work-up of FI [128,129]. The anatomy of the anal sphincters is well visualized facilitating diagnosis of sphincter defects and degenerative changes [56,130,131].

1.9.2   Anal manometry

Anal manometry evaluates the function of the IAS and the EAS, by measuring ARP and ASP respectively. Results depend on technique and equipment [132,133]. Anal pressures are changed with age and gender and the normal ranges are wide [40,134,135]. A set of normal values for women in western Denmark has been established (table 2).

	Women, middle aged	Women, post-menopausal	Women, all
Anal resting pressure (cmH$_2$O)	81 (30-132)	68 (18-118)	82 (29-136)
Anal squeeze pressure (cmH$_2$O)	104 (39-170)	97 (37-157)	109 (43-175)

Table 1.2: Anal manometry reference values. Mean (95% confidence interval)

1.9.3   Rectal sensation

Rectal sensation is examined by filling a highly compliant balloon placed in the rectum. Volumes at first sensation (FS), the desire to defecate (DTD), and the maximal tolerable volume (MTV) are reported. Normal ranges are wide and the percepted volumes increase with age [136]. Large volumes are seen in patients with megarectum and inability to sense indicates neural impairment with rectal hyposensitivity

1.9.4   Pudendal nerve terminal motor latency

The pudendal nerve terminal motor latency is tested with the St. Mark's electrode with a stimulation electrode placed at the tip of the index finger and a recording electrode place at the base (the distance between the electrodes is fixed at 50 mm) [137]. Only the fastest conducting nerve fibers are assessed, and partial neuropathy may produce normal conduction velocities. Latency is recorded bilaterally. The interobserver and intraobserver reproducibility is good [138]. Slow nerve conduction, suggests a damage to the nerve fibers.

1.9.5   Anal sensibility

Anal mucosa electrosensitivity is assessed with a ring electrode placed 1 cm above the anal verge. The current is increased from zero until threshold (first sensation) is reached [139,140].

1.9.6   Colorectal transit time

Total and segmental colonic transit times can be assessed with various radiopaque marker tests [141,142]. Several methods exist. With the most commonly used method the patient ingests a capsule containing 24 markers on day zero. On day five a plain abdominal radiograph is taken. If more than 5 makers are present it suggests constipation. Diffuse scattering indicates colonic inertia (slow transit constipation), while markers piled in the rectosigmoid region suggests an outlet obstruction. The method does not allow quantitative determination of colorectal transit time. In contrast, repeated intake of markers (e.a. on six consecutive days to provide a steady-state) followed by a single abdominal radiograph on the sixth day allows quantitative determination of transit time. The method, however, relies on patients remembering to take the markers.

1.10   METHODS FOR INVESTIGATION OF RECTAL WALL PROPERTIES AND RECTAL MOTILITY

Rectal motility and distensibility play important roles both as part of the continence mechanism and during defecation. Various techniques have been utilized for investigation of these properties. Rectal compliance, defined as dV/dp, has been used to describe both passive rectal wall properties and active rectal distensibility (the sum of both passive viscoelastic wall properties and muscle tone).

1.10.1   Manometric measurements of rectal motility

Rectal motility has been studied with manometric techniques, either perfused or solid state catheters. During contractions that occlude the lumen, the pressure generated by the muscles is fully transmitted, while non-occluding contractions results in measurement of the intraluminal pressure adjacent to the open catheter tip. This makes the technique less useful in non-sphincteric regions. Furthermore, when perfused catheters are used, continuous perfusion may inject a significant volume of water which restricts the length of the investigation. Solid state pressure catheters are better suited for longer recordings.   Furthermore Sasaki et al. demonstrated poor correlation between changes in intraluminal pressure and changes in luminal cross sectional area (CSA) [1].

1.10.2   Pressure-volume measurements with a balloon

A balloon mounted on a catheter is distended at a constant rate, and pressure is measured continuously to assess corresponding mano-volumetric data. This method is hampered by methodological problems, and problems with standardization. Accordingly, very large variations are reported in healthy subject [2-4]. Primarily, the rectum is not a closed ended cylinder, and this model does not take elongation of the balloon in to account. When a standard size probe is used, a mismatch can occur if the rectum is small or large. Regardless of the true wall properties, a large rectum will have a higher measured compliance compared with a smaller rectum [5,6]. Secondly, compliance will be influenced by both changes in smooth muscle tone and contribution of extra-rectal soft tissue. Thirdly, differences in investigational standards affect results. Different shapes and sizes of the balloon, type and speed of distension, distension done with water or air, feeding status, use of enemas, and use of sedation are all factors which influence the compliance measurements.

1.10.3   Pressure-volume measurements with a non-compliant bag

Some of the inherent problems with balloon distension have been overcome by using a large non-compliant bag fixed to a catheter at both ends to avoid any elongation. Krogh et al. compared pressure volume measurements with a compliant balloon with pressure volume measurements with a non-compliant bag and pressure-CSA measurements with impedance planimetry (IP) [7]. Reproducibility for pressure-volume measurement with this noncompliant bag and rectal IP was significantly better than for pressure-volume measurement with a compliant balloon. Accuracy and reproducibility of pressure-volume measurement with a large, noncompliant bag and rectal IP were not statistically different.

1.10.4   Pressure-CSA measurements with impedance planimetry

Rectal IP can be used to determine the potential difference between electrodes mounted on a rectal probe. Given a current I, the potential difference ΔV between two electrodes in a fluid with the conductivity σ, is $\Delta V = I\, d\, \sigma^{-1} CSA^{-1}$, where CSA is the luminal cross sectional area and d is the distance between two detection electrodes. Two excitation electrodes are mounted on a probe, providing a sinusoidal current of 0.1 mA at 10 kHz. Between the excitation electrodes one or more pairs of detection electrodes are mounted. The electrodes are within a non-compliant non-conducting bag made from polyurethane, and filled with 0.9 % saline at 37° C. 20 mm from the distal edge of the rectal bag, a small balloon is mounted on the probe. Simultaneously, the intraluminal rectal pressure and the pressure within the anal canal are measured using perfused catheters within the bag and an anal pressure balloon. The pressure within the bag is controlled by elevation of an open water container. The sample frequency of the system is 10 Hz. The signal conditioning for each channel (CSA, rectal pressure and anal pressure) consists of a sine wave generator, an alternating current generator, an amplifier, a rectifier and a filter. The signal is amplified and converted analogue to digital and forwarded to a computer. Custom made software (Openlab, Noeresundby, Denmark) converts the potentials differences into CSA, visualize data, and stores the data [8,9].

1.10.4.1 Protocols used for impedance planimetric investigations

The system can be used for both isobaric protocols for long recording of rectal motor activity or distension protocols for investigation of rectal wall properties. Rectal IP has been used to study healthy subjects, patients with ulcerative colitis, irradiation damage, and SCI patients [10-13]. When measurements are done using a distension protocol, a number of repeated distensions are needed to obtain reproducible responses. Otherwise, the pressure-CSA relation may depend upon the loading history.

1.10.4.2 Sources of error with impedance planimetry

There are a number of potential sources of error related to IP measurements. The probe can be dislocated to an eccentric position in the rectum. This can be avoided by using a large bag that unfolds. The position of the bag has been tested with transabdominal ultrasonography in few occasions and demonstrated to be placed centrally and parallel to the rectal wall [14]. A change in temperature will affect fluid conductivity and cause an error on CSA measurements of up to 3% per degree [15]. The slope of the wall between the detection electrodes constitutes a theoretical error. However, the rectum is fairly straight so it may be neglected. Sample frequency should be much higher than frequency of measured contractions to overcome aliasing. This is no concern during measurements in the rectum since contraction frequency is not above 10 min^{-1} and the sample frequency is 10 Hz. Resistance to flow of water within the probe must be considered when distension is performed or phasic contractions are present. This is particularly a concern when distension is performed with elevation of an open water container. The error related to water flow can be minimized by making the tube for infusion as short as possible and with as large a diameter as possible. Ultimately, controlling distension with a pump will overcome the problem. Hysteresis within the system, i.e. different pressure-CSA relation during inflation and deflation, should also be considered.

1.10.5   Probes for recording of rectal phasic activity and wall properties during distension

Two different probes were used to investigate phasic rectal motor activity and rectal wall properties during distension. For recording of phasic rectal motor activity, a probe with five pairs of detection electrodes was used. It allowed description of spatial and temporal distribution of phasic rectal contractions and characterization of amplitude and direction. The equipment and the protocol for measurement of rectal phasic motor activity have previously been described in detail [16]. The electrodes were mounted on an 18-cm-long probe with an outer diameter of 9 mm (TensioMED, Hornslet, Denmark). The probe and the electrodes were covered with a non-compliant flaccid bag (14 cm long, diameter 8 cm, 450 ml capacity) made of 50 µm thick polyurethane. Through a central channel with several side holes the bag could be filled with a conductive fluid. The excitation electrodes were separated by 125 mm and the detection electrodes were placed between the excitation electrodes. The inter-electrode distance in a pair of electrodes was 2 mm and the distance between each pair was 20 mm. A pressure transducer connected to the bag with a thin catheter inside the rectal balloon measured the rectal pressure. Pressure was controlled by elevation of a water container. Distal to the rectal balloon, an anal balloon was mounted. It was filled with 4 ml saline and connected with a transducer which measured anal pressure. The *in vitro* accuracy of CSA measurements with this probe was 2.3-6.7 percent, and the correct placement of the probe in the rectum was verified by ultrasound [16,17].

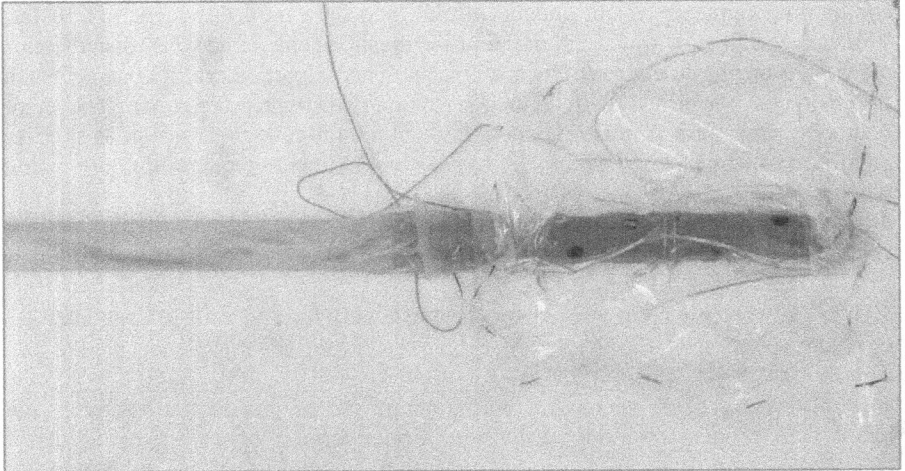

Figure 1.3: The IP probe used for distension.

For the distension protocol a probe with one pair of detection electrodes was used (fig A1). Otherwise the probe was constructed similarly. The diameter of the bag was 90 mm. The distance between the excitation electrodes was 62 mm and the two detection electrodes 2 mm apart, right between the excitation electrodes. Limiting the number of detection electrodes provided a flow rate of 13-17 ml s^{-1} during filling of the rectal balloon, which allowed an immediate response to distension.

1.10.6   Calibration procedure

As calibration curves for IP are non-linear, a multipoint calibration is made before measurements. We used 15 tubes with CSAs in the range 281-4322 mm^2. The curve was linearized by stepwise interpolation between the calibration measurements (fig A2). The calibration constant depends on current intensity, fluid conductivity, temperature, and distance between the excitation electrodes.

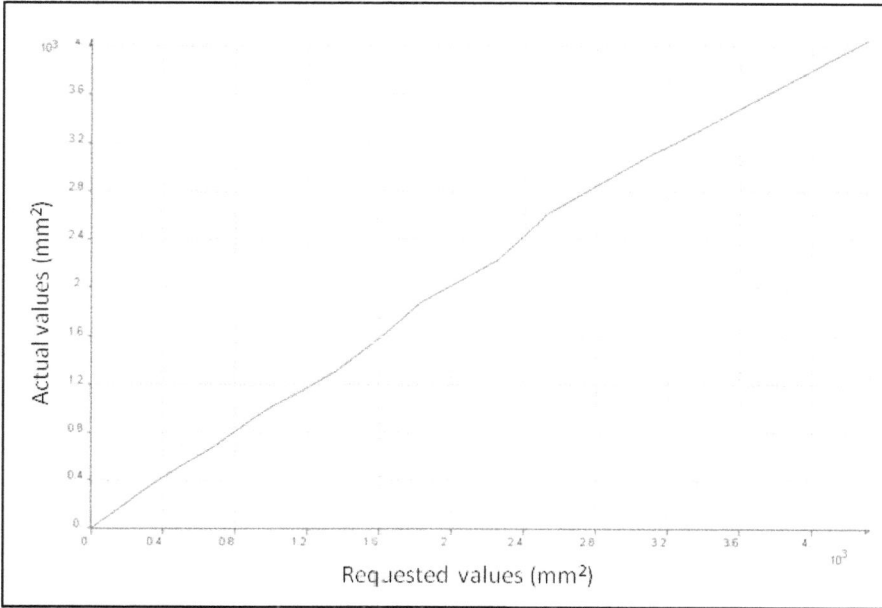

Figure 1.4: Calibration curve. On the x-axis are the values of the plastic tubes and on the y-axis are the measured cross sectional areas.

1.10.7   Validity of the impedance planimetry probe used for distension

Before undertaking experiments, the probe was tested *in vitro* using plastic tubes with known cross sectional areas ranging from 3.30 to 43.22 cm^2 thus covering the expected range of measurements. On two different days, 110 measurements were done in eleven tubes. Accuracy (validity of measurements) was reported as the difference (numeric value) between the measured cross sectional area and the actual cross sectional area of the plastic tube. Precision (reproducibility of measurements) was reported as the difference between the measured cross sectional area and the average of measured cross sectional areas.

Validity and reproducibility was dependent on the size of CSA measurements. The accuracy decreased when the size of CSA increased (fig 1.5). With CSAs below 30 cm^2 the maximal error was approximately 400 mm^2. The relative accuracy did not change with the size of CSA.

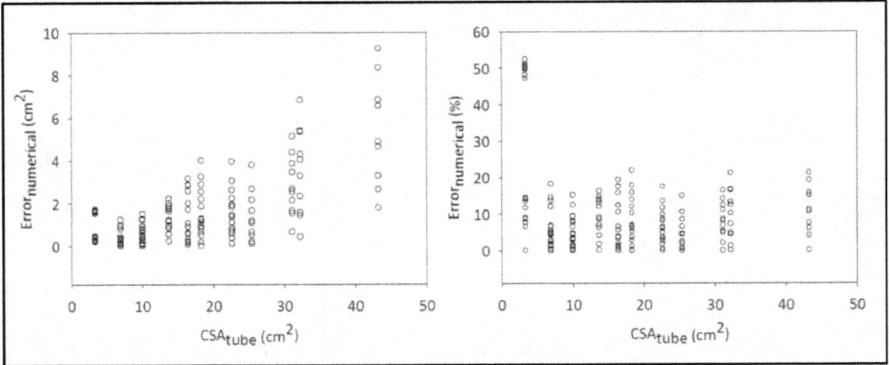

Figure 1.5: Validity of IP measurements. The error increases when CSA exceeds 30 cm^2.

Data are given in table 1.3. There was a notable difference in accuracy between day 1 and day 2. Sources of error related to IP measurements have already been described. There was no systematic difference in the procedure or the experimental setup between measurements from day 1 and day 2. The overall error was 7.3%.

	Accuracy, all	Accuracy day 1	Accuracy day 2
Mean error (cm^2)	1.2	0.8	1.8
CI95%	1.1-1.3	0.7-0.9	1.7-0.8
Range (cm^2)	0 to 9.3	0 to 8.4	0.1 to 9.3
Mean error (CI95%)	7.3	5	11
CI95%	0-14	0-12	6-16
Range (%)	0 to 52	0 to 19	1 to 52

Table 1.3: In vitro validity of IP measurements.

The precision also decreased when measuring large CSAs (fig 1.6).The relative precision did not change with the size of CSA. Data are given in table 1.4.

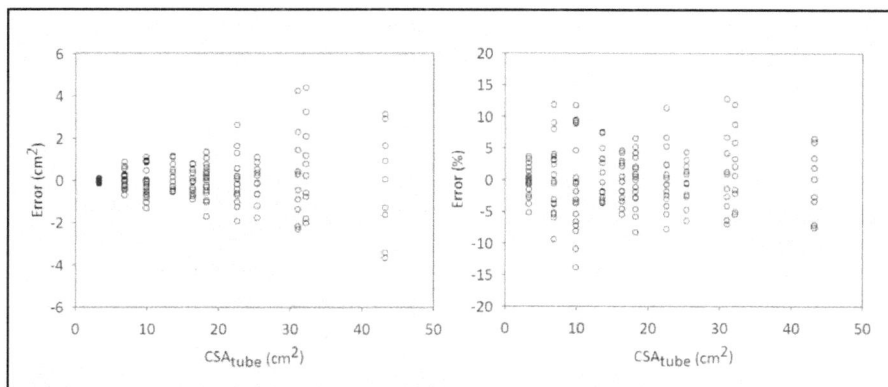

Figure 1.6: Reproducibility of IP measurements. CSA measurements below 30 cm² are reproducible. When CSA exceeds 30 cm², the variation increases.

	Precision, all	Precision day 1	Precision day 2
CI95% (cm²)	±2.0	±2.0	±2.2
Range (cm²)	-3.7 to 4.4	-3.4 to 4.3	-3.7 to 4.4
CI95% (%)	±9	±9	±8
Range (%)	-14 to 13	-14 to 13	-9 to 12

Table 1.4: In vitro reproducibility of IP measurements.

1.11   CURRENT TREATMENT OPTIONS FOR FI

Management of FI is stepwise with successively more invasive interventions. Patients suffering from severe FI refractory to conservative management can be offered various surgical treatments including sacral nerve stimulation, appendicostomy, various sphincter reconstructive procedures or ultimately a stoma [6,155-157]. Generally, treatment of FI is empirically based and not much hard evidence exists to guide clinical practice [158-163]. The following includes an overview of the options for patients with idiopathic FI and SCI patients with FI.

1.11.1   Conservative treatment

Often simple conservative measures are effective. Lifestyle alteration including physical exercise and weight loss are effective [164,165]. Dietary advice to promote intake of sufficient fluid and fibers and advice about bowel habits can be useful [166,167]. Pharmacological interventions include bulking agents and agents that suppress propulsive activity (loperamide, diphenoxylate, codeine, phosphate, bile acid binders).   Minor FI can be managed with pads or anal plugs.

Suppositories and mini enemas can be used to keep the rectum empty to avoid incontinence [168].

1.11.2   Biofeedback

The aim of biofeedback training is to improve the patient's perception of rectal sensation and also to systematically exercise the external sphincter and the pelvic floor (levator ani muscles) [169]. The treatment is best suited for patients with partial damage of the pelvic nerves but can also benefit patients with small sphincter defects [170,171]. Patients commit to a 6-8 weeks training program, where after the effect is reviewed. There is disagreement concerning indications and standardization, and only a limited number of well-conducted studies with clear outcome parameters exist [172,173]. It remains to be determined if treatment is more effective than placebo [172,174].

1.11.3   Transanal irrigation

The principle of transanal irrigation is to avoid fecal leakage by systematic scheduled irrigation of the rectum and the left colon. Irrigation is performed daily or every other day. A number of catheters exist including a special catheter with a balloon to keep it the rectum while the enema is administered (enema continence catheter). Usually, tap water is used. The irrigation volume needed varies. A study using scintigraphy, demonstrated an almost complete emptying of the rectosigmoid and descending colon in SCI patients and patients with idiopathic FI [175]. Transanal irrigation has also been demonstrated to reduce symptoms in patients with NBD and idiopathic FI [176]. In SCI patients, it is less expensive and more effective than conservative treatment. However, long term success rate is only 35% after three years [177,178].

1.11.4   Surgery (Malone, levatorplasty, postanal repair, neo-sphincter, magnet-sphincter)

Malone et al described a method of tunneling the reversed appendix in order to secure continence [155]. Other procedures inspired by Malone are in-situ appendicostomy and tubularized ileal or cecal conduits [179-186]. ACE improves bowel function, social function and quality of life in patients with FI caused by SCI and in patients with idiopathic FI [187]. One study found that antegrade irrigation was superior to sacral anterior root stimulation or formation of a colostomy or an ileostomy in patients with NBD [188]. In selected cases the ACE has also been combined with a colostomy [189].

Procedures targeting the structure of the pelvic floor (anterior levatorplasty and postanal repair) have been applied in patients with FI [190]. Results have not been convincing with continuing high FI scores reported postoperatively [191-193]. The artificial anal sphincter and dynamic graciloplasty are extensive procedures which can be effective in special cases [194-196]. Also, a magnetic sphincter has been constructed to dynamically sustain sufficient anal pressure and allow defecation [197]. These artificial sphincter devices would not be an option in SCI patients, since NBD most often includes both FI and constipation. The ultimate treatment is fecal diversion through a stoma, which can be the best option to gain control and improve quality of life in both idiopathic and neurogenic patients [198,199].

1.12   ELECTRICAL STIMULATION IN THE MANAGEMENT OF FI

1.12.1   Sacral nerve stimulation

Sacral nerve stimulation using the Interstim therapy system (Medtronic, USA) was introduced for idiopathic FI in 1995 [18]. An electrode is placed through a sacral foramen S2-S4 (preferably S3) to stimulate sacral nerves. The procedure comprises two stages; a three week test period serves to evaluate the efficacy of treatment and a 50% improvement of symptom score is used as the cut off for progression to the second stage; the implantation of the permanent stimulator [19]. For the test period, one or more test stimulation leads or alternatively a permanent lead with an extension cable is used and connected to an external pulse generator. Usually, implantation of the permanent electrode is unilateral, guided by the response using the test lead or leads, if more leads have been used during the test period. The permanent pulse generator is placed in a gluteal pocket, where it is accessible for programming by external telemetry and replacement. Stimulation parameters are adopted from experiences with treatment of urinary symptoms (pulse with of 210 μs, a frequency of 15 Hz, and the amplitude set individually usually in the range between 0.1 V to 10 V).

The result from the test period has a high predictive value for subsequent successful permanent implantation. Even so, no specific demographic or pathophysiological predictors for a positive test in patients with FI have been identified. SNS has proven effective in reducing episodes of FI and improving quality of life in the short term [19-24]. Furthermore, the effect appears to be maintained at both medium (six years) and long-term follow-up (ten years and fourteen years) [25-27]. Even so, some loss of efficacy and adverse effects (pain/discomfort) are more likely to produce suboptimal outcome at later stages [28].
Initially, indications were limited to patients with deficient function of a morphologic intact sphincter [19,29]. Now SNS has been successfully applied to patients with sphincter injuries, inflammatory bowel disease, partial SCI, colorectal resection, rectal prolapse, sequelae from chemo-radiation, constipation, and irritable bowel syndrome [30-36].

The mechanism of action still remains largely unsolved, but likely better understanding will aid to define indications and optimize stimulations parameters.

1.12.2   Peripheral electrical nerve stimulation

Peripheral electrical stimulation for the management of FI can include both nerve stimulation and transanal stimulation of pelvic floor and sphincter muscles. With nerve stimulation, an electrical pulse is applied to an afferent neural pathway and through synaptic transmission activates or inhibit neural pathways innervating the bowel, pelvic floor and anal sphincters. The afferent transmission is conducted via thick myelinized Aα-fibers. They can be activated at an amplitude below the threshold for un-myelinized nociceptive Aδ-fibers and c-fibers, and this is advantageous in non-neurogenic patients with preserved sensibility. Generally, optimal stimulation parameters should depolarize the nerve at a minimum charge per pulse to minimize energy consumption and avoid discomfort for the patient. One study investigated sensory and motor thresholds at pulse duration of different lengths, and a pulse duration of 0.2 ms was had the optimal pulse energy - pulse duration relationship [37]. Stimulation frequency can also be changed with use of high frequencies limited by increased power consumption, nerve damage, pain and refractory period of the nerve [38,39].

Management of FI by both DGN stimulation and PTNS stimulation is used. The pudendal nerve is a mixed sensory-motor nerve originating from S2-S4. The dorsal genital branch is easily accessible carrying afferent fibers to the sacral roots. The posterior tibial nerve (L4-S3) is also a mixed sensory-motor nerve with afferent pathways going to the lumbosacral dorsal roots. It can be accessed above the medial malleolus [40]. Efferent outflow from Onuf's nucleus (primarily located at the S2 level) goes through somatic fibers innervating the EAS and makes the striated muscle amenable for training through electrically stimulated repetitive contractions. Autonomic outflow from the lumbosacral spinal cord constitutes the efferent pathway to the bowel and the IAS. Both the sacral part of the parasympathetic nervous system (originating from S2-S4) and sympathetic nerves (originating mainly from L1-L3) could theoretically be stimulated causing a change of the colonic and anorectal function.

With direct stimulation of the muscles of the pelvic floor and the external sphincter, the aim is to induce repetitive contractions thereby improving muscle strenght [41-43]. This has mostly been applied as transanal stimulation in conjunction with biofeedback training.

Peripheral electrical nerve stimulation has been used for management of patients with voiding dysfunction or urinary incontinence. Stimulation of the posterior tibial nerve has been successful in patients with overactive bladder and detrusor hypocontractility [44-50]. Acute DGN stimulation has been demonstrated to inhibit contractions in patients with idiopathic detrusor overactivity and in patients with neurogenic detrusor overactivity. Acute stimulation was able to inhibit bladder contractions and decrease the filling pressure [51-56].

1.12.2.1 Dorsal genital nerve stimulation

The acute effect of DGN stimulation was invetigated by *Chung et al.* [57]. Rectal compliance was studied in patients suffering from supra sacral SCI. A rectal barostat balloon was filled both during and without electrical stimulation and rectal compliance was found to increase with stimulation. Different stimulation frequencies were used (0.2 Hz, 2 Hz, 20 Hz) and at each frequency compliance was significantly higher compared with unstimulated filling. With stimulation at 20 Hz the highest compliance was seen.

Two studies have investigated the effect of DGN stimulation on FI symptoms and anal sphincter function. *Binnie et al.* treated eight women with socially incapacitating FI arising from pudendal neuropathy [58]. Pulse width was 100 μs, frequency 1 Hz, and mean amplitude was 135 (sd: 15) V. Stimulation was applied using saline soaked plaster electrodes and the patients were treated for 5 minutes 3 times a day for eight weeks. The mean resting pressure in the anal canal and the reflex and voluntary pressure to coughing and squeezing of the EAS all increased significantly after therapy. Seven of eight patients experienced an improvement gaining continence for feces and flatus at the end of the treatment. A study by *Frizelle et al.* examined 42 (39 females) patients presenting with FI from unspecified reasons [59]. They underwent ano-rectal physiology testing and symptom scoring with the Cleveland Fecal Incontinence Score before and after eight weeks treatment with stimulation for five minutes 3 times a day. Pulse width was 100 μs, frequency was 2 Hz, and the amplitude was individually titrated in the range from 30 to 90 V. After intervention their maximum resting pressure, maximum squeeze pressure and maximum tolerable volume all were significantly higher. There was also a significant reduction in mean incontinence score from 9.3 to 6 and there was a significant reduction in the number of patients with incontinence to solid or liquid stools more than once weekly after the threatment.

1.12.2.2 Stimulation of the posterior tibial nerve in management of FI

Posterior tibial nerve stimulation (PTNS) was first indicated exclusively for bladder dysfunction. Many of these patients suffer from both urinary incontinence and FI (double incontinence) and some also experienced improvement of FI symptoms. Therefore, PTNS was investigated further for the management of FI. Several studies investigating PTNS have been published [60-68]. PTNS has been evaluated in patients with FI due to, sphincter defects resulting from obstetric trauma or anorectal surgery, inflammatory bowel disease, and idiopathic etiology. All studies were case series including from two to 31 patients.

Stimulation has been done, either using self-adhesive surface electrodes for transcutaneous stimulation or by needle electrodes for percutaneous stimulation. Self-adhesive electrodes were placed on the skin behind the medial malleolus. Percutaneous stimulation was done according to Stoller et al. with a needle inserted 3-4 cm cephalad to the medial malleolus [69]. A ground surface electrode is placed on the ipsilateral leg near the arch of the foot. In all studies with PTNS, pulse width was 0.2 ms (not reported by Findlay et al.) and the frequency was 10 or 20 Hz. The amplitude setting differs from below motor threshold to maximal tolerable current but generally stimulation amplitude is below 10 mA. Different treatment protocols have been applied ranging from 4 weeks to 12 weeks treatment with scheduled stimulation sessions from daily to every third to fourth day. Maintenance therapy following initial treatment has also been investigated.

The outcome of the treatment has been evaluated with bowel habit diaries before and after treatment in three studies reporting a decrease in the number of incontinence episodes after treatment. The ability to defer defecation was reported to improve after treatment in two studies. Wexner FI score was reported by seven authors before and after treatment, with significant improvement in all but one study which included IBD patients. The fecal incontinence quality of life score (Rockwood score) was improved in four studies, however only the lifestyle score was improved in the study be Findley et al. [60-68]. The SF-36 health survey was improved in one study [67]. The Hospital Anxiety-and-Depression score assessed by Finley et al. did not improve after treatment [66]. De la Portilla et al. reported improvement of 10 out of 16 patients, and Boyle et al. reported improvement in 21 out of 31 patients; both studies used weekly stimulation for 12 weeks [63,65]. Furthermore, after the initial treatment, maintenance therapy with stimulatin every other week appears to be sufficient to sustain the reduction in symptoms [63]. Results from anal manometry done in three studies before and after treatment are ambiguous [61-63]. Rectal sensation investigated with a rectometrogram was found to be improved by Shafik et al., while Portilla et al. did not demonstrated any change in first sensation or urgency during rectal balloon distension [60,63].

1.12.3   Other stimulation methods

1.12.3.1 Transanal stimulation in the management of FI

Transanal electrical stimulation is used for FI either as sole treatment or as a part of augmented biofeedback [70]. Indications are partial sphincter injury after obstetric trauma or neurogenic FI due to damage (stretch, compression, degenerative) of the pudendal nerve. The stimulus is delivered via an anal electrode applied with a conductive lubricant. The effect of transanal electrical stimulation has mostly been evaluated with anal manometry and FI severity scores. Several case series and retrospective studies have reported electrical stimulation to benefit fecal incontinent patients while others have not been able to demonstrate any gain from stimulation [71-73,73-75]. Recently, a Cochrane review indentified four randomized controlled trials [70,76-78]. Transanal stimulation was compared with other treatments or sham stimulation. The review concluded that present knowledge is insufficient to draw reliable conclusions on the effect of transanal electrical stimulation in the management of FI [79].

It is hypothesized that electrical stimulation improves strength, speed or endurance of the voluntarily controlled striated sphincter muscle. There are some suggestions as to how this is achieved. Dependent on the stimulation frequency, electrical stimulation can transform fast twitch muscle fibers into slow twitch muscle fibers thereby improving endurance of muscle function [41]. Electrical stimulation has also been shown to increase capillary density which provides better oxygenation to the muscles fibers [42]. Another effect may be improved anorectal sensation and awareness of sphincter function [80]. Optimal stimulation parameters have not been established. Testing has been carried out using a wide range of setups, stimulation parameters, and stimulation protocols. Most clinical studies have utilized stimulation with trains of pulses (200 to 300 µs) at a frequency from 20-50 Hz sufficient to elicit muscle contraction. Nie et al. tested different protocols with acute long pulse (50-300 ms) stimulation at a low frequency (0.25 to 0.5 Hz) in dogs and found that sphincter response was dependent on the amplitude, and not subject to muscle fatigue during prolonged stimulation for 20 min [81]. It was also demonstrated that α-adrenergic but not cholinergic pathways may be involved, as the excitatory effect of stimulation was inhibited by phentolamine (α-adrenergic antagonist) but not by atropine (acethylcholine receptor antagonist) [82].

1.12.3.2 Transvaginal stimulation in the management of FI

Transvaginal stimulation has also been proposed, as an alternative method. Song et al. investigated the effect on rectal tone and sphincter pressure with different modes of transvaginal stimulation [83]. Stimulation parameters were adopted from Nie et al. [81]. The stimulation was composed of long pulses (pulse width: 500 ms, pulse amplitude: 6 mA, frequency: 20 cycles per minute), trains of long pulses (train on-time of 2 s and off-time of 3 s, pulse width: 4 ms, amplitude 6 mA, and pulse frequency: 40 Hz), or trains of short pulses (same as trains of long pulses except a pulse width of 0.3 ms). A pair of ring electrodes (2 cm apart) were used and mounted on a catheter. Stimulation with long pulses or trains of long pulses but not trains of short pulses significantly decreased rectal tone and increased anal sphincter pressure, but rectal compliance was not affected. The proposed mode of action was a modulation of vaginal structures, adjacent ligaments, and the rectovaginal fascia, thereby strengthening the pelvic floor and supporting anorectal function. Infusion of guanethidine (adrenergic blocker) abolished the inhibitory effect of stimulation on rectal tone suggesting some involvement of parasympathetic pathways.

1.12.3.3 Magnetic stimulation

The use of magnetic stimulation to change colonic motility has been investigated with a magnetic coil placed over the S3 roots [84]. In a randomized setup with magnetic stimulation and sham stimulation, healthy subjects were given bisacodyl to promote colonic activity. Motility was measured using a manometry catheter. The appearance of high-pressure contractions propagated or not (HAC/HAPC) provoked by Bisacodyl instillation was significantly delayed during stimulation. The perception of urgency tended to be lower with stimulation following Bisacodyl instillation. After Bisacodyl administration, the catheter was expulsed significantly slower during stimulation compared sham with stimulation.

REFERENCES

[1]   Whitehead WE, Wald A, Diamant NE, Enck P, Pemberton JH, Rao SS. Functional disorders of the anus and rectum. Gut 1999;45 Suppl 2:II55-9.

[2]   Tanagho EA, Schmidt RA. Bladder pacemaker: scientific basis and clinical future. Urology 1982;20:614-9.

[3]   Schmidt RA. Advances in genitourinary neurostimulation. Neurosurgery 1986;19:1041-4.

[4]   Tanagho EA, Schmidt RA. Electrical stimulation in the clinical management of the neurogenic bladder. J Urol 1988;140:1331-9.

[5]   Schmidt RA, Senn E, Tanagho EA. Functional evaluation of sacral nerve root integrity. Report of a technique. Urology 1990;35:388-92.

[6]   Matzel KE, Stadelmaier U, Hohenfellner M, Gall FP. Electrical stimulation of sacral spinal nerves for treatment of faecal incontinence. Lancet 1995;346:1124-7.

[7]   Andreasen E, Bierring F, Rostgaard J. De indre organers anatomi. 8th edition ed. Copenhagen: Munksgaard, 1995.

[8]   Rhoades RA, Tanner GA. Medical Physiology. In: Anonymous , USA: Little, Brown; 1995, p. 505-529.

[9]   Habr-Gama A, Jorge JMN. Anatomy and Embryology of the Colon, Rectum, and AnusinThe ASCRS Textbook of Colon and Rectal Surgery (Editors: Wolff, B.G.; Fleshman, J.W.; Beck, D.E.; Pemberton, J.H.; Wexner, S.D.). 2007:pp 1-22-22.

[10]   Wood BA, Kelly AJ. Anatomy of the anal sphincters and pelvic floor in Coloproctology and the pelvic floor (editors: Henry, M.M; Swash, M.). 1992:pp 3-19.

[11]   Milligan ETC, Morgan CN. Surgical anatomy of the anal canal: with special reference to anorectal fistulae. Lancet 1934;2:1150-6.

[12]   Benarroch EE. Enteric nervous system: functional organization and neurologic implications. Neurology 2007;69:1953-7.

[13]   Costa M, Brookes SJ, Hennig GW. Anatomy and physiology of the enteric nervous system. Gut 2000;47 Suppl 4:iv15,9; discussion iv26.

[14]   Brodal P. Sentralnervesystemet, Bygning og Funksjon. 2nd. ed. Otta, Norway: AiT Enger AS, 1995.

[15] Erspamer V, Asero B. Identification of enteramine, the specific hormone of the enterochromaffin cell system, as 5-hydroxytryptamine. Nature 1952;169:800-1.

[16] Spiller RC. Postinfectious irritable bowel syndrome. Gastroenterology 2003;124:1662-71.

[17] Nemeth PR, Ort CA, Wood JD. Intracellular study of effects of histamine on electrical behaviour of myenteric neurones in guinea-pig small intestine. J Physiol 1984;355:411-25.

[18] Liu S, Xia Y, Hu H, Ren J. Gao C, Wood JD. Histamine H3 receptor-mediated suppression of inhibitory synaptic transmission in the submucous plexus of guinea-pig small intestine. Eur J Pharmacol 2000;397:49-54.

[19] Kreis ME, Jiang W, Kirkup AJ, Grundy D. Cosensitivity of vagal mucosal afferents to histamine and 5-HT in the rat jejunum. Am J Physiol Gastrointest Liver Physiol 2002;283:G612-7.

[20] Sanders KM. A case for interstitial cells of Cajal as pacemakers and mediators of neurotransmission in the gastrointestinal tract. Gastroenterology 1996;111:492-515.

[21] Fleckenstein P. Migrating electrical spike activity in the fasting human small intestine. Am J Dig Dis 1978;23:769-75.

[22] Fleckenstein P, Oigaard A. Electrical spike activity in the human small intestine. A multiple electrode study of fasting diurnal variations. Am J Dig Dis 1978;23:776-80.

[23] Snape WJ,Jr, Carlson GM, Cohen S. Human colonic myoelectric activity in response to prostigmin and the gastrointestinal hormones. Am J Dig Dis 1977;22:881-7.

[24] Bueno L, Fioramonti J, Ruckebusch Y, Frexinos J, Coulom P. Evaluation of colonic myoelectrical activity in health and functional disorders. Gut 1980;21:480-5.

[25] Taylor I, Duthie HL, Smallwood R, Linkens D. Large bowel myoelectrical activity in man. Gut 1975;16:808-14.

[26] Szurszewski JH. A migrating electric complex of canine small intestine. Am J Physiol 1969;217:1757-63.

[27] Vantrappen G, Janssens J, Hellemans J, Ghoos Y. The interdigestive motor complex of normal subjects and patients with bacterial overgrowth of the small intestine. J Clin Invest 1977;59:1158-66.

[28] Kellow JE, Borody TJ, Phillips SF, Tucker RL, Haddad AC. Human interdigestive motility: variations in patterns from esophagus to colon. Gastroenterology 1986;91:386-95.

[29] Husebye E, Skar V, Aalen OO, Osnes M. Digital ambulatory manometry of the small intestine in healthy adults. Estimates of variation within and between individuals and statistical management of incomplete MMC periods. Dig Dis Sci 1990;35:1057-65.

[30] Husebye E, Engedal K. The patterns of motility are maintained in the human small intestine throughout the process of aging. Scand J Gastroenterol 1992;27:397-404.

[31] Hansen MB. Small intestinal manometry. Physiol Res 2002;51:541-56.

[32] Husebye E. The patterns of small bowel motility: physiology and implications in organic disease and functional disorders. Neurogastroenterol Motil 1999;11:141-6.

[33] Herbst F, Kamm MA, Morris GP, Britton K, Woloszko J, Nicholls RJ. Gastrointestinal transit and prolonged ambulatory colonic motility in health and faecal incontinence. Gut 1997;41:381-9.

[34] Rao SS, Welcher K. Periodic rectal motor activity: the intrinsic colonic gatekeeper? Am J Gastroenterol 1996;91:890-7.

[35] Kumar D, Williams NS, Waldron D, Wingate DL. Prolonged manometric recording of anorectal motor activity in ambulant human subjects: evidence of periodic activity. Gut 1989;30:1007-11.

[36] Orkin BA, Hanson RB, Kelly KA. The rectal motor complex. Journal of Gastrointestinal Motility 1989;1:5-8.

[37] Prior A, Fearn UJ, Read NW. Intermittent rectal motor activity: a rectal motor complex? Gut 1991;32:1360-3.

[38] Auwerda JJ, Bac DJ, Schouten WR. Circadian rhythm of rectal motor complexes. Dis Colon Rectum 2001;44:1328-32.

[39] Lestar B, Penninckx F, Kerremans R. The composition of anal basal pressure. An in vivo and in vitro study in man. Int J Colorectal Dis 1989;4:118-22.

[40] Ryhammer AM, Laurberg S, Hermann AP. Test-retest repeatability of anorectal physiology tests in healthy volunteers. Dis Colon Rectum 1997;40:287-92.

[41] Hancock BD. Measurement of anal pressure and motility. Gut 1976;17:645-51.

[42] Sorensen SM, Gregersen H, Sorensen S, Djurhuus JC. Spontaneous anorectal pressure activity. Evidence of internal anal sphincter contractions in response to rectal pressure waves. Scand J Gastroenterol 1989;24:115-200.

[43] Enck P, Eggers E, Koletzko S, Erckenbrecht JF. Spontaneous variation of anal "resting" pressure in healthy humans. Am J Physiol 1991;261:G823-6.

[44] Denny-Brown D, Robertson EG. An investigation of the nervous control of defecation. 1935;58:256-310.

[45] van Duijvendijk P, Slors F, Taat CW, Heisterkamp SH, Obertop H, Boeckxstaens GE. A prospective evaluation of anorectal function after total mesorectal excision in patients with a rectal carcinoma. Surgery 2003;133:56-65.

[46] Malcolm A, Camilleri M. Coloanal motor coordination in association with high-amplitude colonic contractions after pharmacological stimulation. Am J Gastroenterol 2000;95:715-9.

[47] Baeten C, Kuijpers HC. Incontinence in The ASCRS Textbook of Colon and Rectal Surgery (Editors: Wolff, B.G.; Fleshman, J.W.; Beck, D.E.; Pemberton, J.H.; Wexner, S.D.). 2007:pp 653-664.

[48] Whitehead WE, Borrud L, Goode PS, Meikle S, Mueller ER, Tuteja A et al. Fecal incontinence in US adults: epidemiology and risk factors. Gastroenterology 2009;137:5_2,7, 517.e1-2.

[49] Rey E, Choung RS, Schleck CD, Zinsmeister AR, Locke GR,3rd, Talley NJ. Onset and risk factors for fecal incontinence in a US community. Am J Gastroenterol 2010;105:412-9.

[50] Walter S, Hjortswang H, Holmgren K, Hallbook O. Association between bowel symptoms, symptom severity, and quality of life in Swedish patients with fecal incontinence. Scand J Gastroenterol 2010.

[51] Markland AD, Greer WJ, Vogt A, Redden DT, Goode PS, Burgio KL et al. Factors impacting quality of life in women with fecal incontinence. Dis Colon Rectum 2010;53:1148-54.

[52] Malmstrom TK, Andresen EM, Wolinsky FD, Schootman M, Miller JP, Miller DK. Urinary and fecal incontinence and quality of life in African Americans. J Am Geriatr Soc 2010;58:1941-5.

[53] Madoff RD, Parker SC, Varma MG, Lowry AC. Faecal incontinence in adults. Lancet 2004;364:621-32.

[54] Faltin DL, Sangalli MR, Roche B, Floris L, Boulvain M, Weil A. Does a second delivery increase the risk of anal incontinence? BJOG 2001;108:684-8.

[55] MacArthur C, Glazener CM, Wilson PD, Herbison GP, Gee H, Lang GD et al. Obstetric practice and faecal incontinence three months after delivery. BJOG 2001;108:678-83.

[56] Sultan AH, Kamm MA, Hudson CN, Bartram CI. Third degree obstetric anal sphincter tears: risk factors and outcome of primary repair. BMJ 1994;308:887-91.

[57] Varma A, Gunn J, Gardiner A, Lindow SW, Duthie GS. Obstetric anal sphincter injury: prospective evaluation of incidence. Dis Colon Rectum 1999;42:1537-43.

[58] Ryhammer AM, Laurberg S, Hermann AP. Long-term effect of vaginal deliveries on anorectal function in normal perimenopausal women. Dis Colon Rectum 1996;39:852-9.

[59] Sultan AH, Kamm MA, Hudson CN, Thomas JM, Bartram CI. Anal-sphincter disruption during vaginal delivery. N Engl J Med 1993;329:1905-11.

[60] Snooks SJ, Setchell M, Swash M, Henry MM. Injury to innervation of pelvic floor sphincter musculature in childbirth. Lancet 1984;2:546-50.

[61] Garcia-Aguilar J, Belmonte C, Wong WD, Lowry AC, Madoff RD. Open vs. closed sphincterotomy for chronic anal fissure: long-term results. Dis Colon Rectum 1996;39:440-3.

[62] Nyam DC, Pemberton JH. Long-term results of lateral internal sphincterotomy for chronic anal fissure with particular reference to incidence of fecal incontinence. Dis Colon Rectum 1999;42:1306-10.

[63] MacIntyre IM, Balfour TW. Results of the Lord non-operative treatment for haemorrhoids. Lancet 1972;1:1094-5.

[64] Read MG, Read NW, Haynes WG, Donnelly TC, Johnson AG. A prospective study of the effect of haemorrhoidectomy on sphincter function and faecal continence. Br J Surg 1982;69:396-8.

[65] Otto IC, Ito K, Ye C, Hibi K, Kasai Y, Akiyama S et al. Causes of rectal incontinence after sphincter-preserving operations for rectal cancer. Dis Colon Rectum 1996;39:1423-7.

[66] Varma MG, Brown JS, Creasman JM, Thom DH, Van Den Eeden SK, Beattie MS et al. Fecal incontinence in females older than aged 40 years: who is at risk? Dis Colon Rectum 2006;49:841-5.

[67] O'Keefe EA, Talley NJ, Zinsmeister AR, Jacobsen SJ. Bowel disorders impair functional status and quality of life in the elderly: a population-based study. J Gerontol A Biol Sci Med Sci 1995;50:M184-9.

[68] Kangas E, Hiltunen KM, Matikainen M. Anorectal function in Crohn's disease. Ann Chir Gynaecol 1992;81:43-7.

[69] Lundby L, Krogh K, Jensen VJ, Gandrup P, Qvist N, Overgaard J et al. Long-term anorectal dysfunction after postoperative radiotherapy for rectal

cancer. Dis Colon Rectum 2005;48:1343,9; discussion 1349-52; author reply 1352.

[70] Gladman MA, Aziz Q, Scott SM, Williams NS, Lunniss PJ. Rectal hyposensitivity: pathophysiological mechanisms. Neurogastroenterology and Motility 2009;21:511-2.

[71] Bharucha AE, Zinsmeister AR, Locke GR, Seide BM, McKeon K, Schleck CD et al. Risk factors for fecal incontinence: a population-based study in women. Am J Gastroenterol 2006;101:1305-12.

[72] Ilnyckyj A. Prevalence of idiopathic fecal incontinence in a community-based sample. Can J Gastroenterol 2010;24:251-4.

[73] McHugh SM, Diamant NE. Effect of age, gender, and parity on anal canal pressures. Contribution of impaired anal sphincter function to fecal incontinence. Dig Dis Sci 1987;32:726-3.

[74] Mitrani C, Chun A, Desautels S, Wald A. Anorectal manometric characteristics in men and women with idiopathic fecal incontinence. J Clin Gastroenterol 1998;26:175-8.

[75] Stojkovic SG, Balfour L, Burke D, Finan PJ, Sagar PM. Role of resting pressure gradient in the investigation of idiopathic fecal incontinence. Dis Colon Rectum 2002;45:668-73.

[76] Rasmussen OO, Ronholt C, Alstrup N, Christiansen J. Anorectal pressure gradient and rectal compliance in fecal incontinence. Int J Colorectal Dis 1998;13:157-9.

[77] Rao SS, Ozturk R, Stessman M. Investigation of the pathophysiology of fecal seepage. Am J Gastroenterol 2004;99:2204-9.

[78] Chang HS, Myung SJ, Yang SK, Jung HY, Kim TH, Yoon IJ et al. Effect of electrical stimulation in constipated patients with impaired rectal sensation. Int J Colorectal Dis 2003;18:433-8.

[79] Ricciardi R, Mellgren AF, Madoff RD, Baxter NN, Karulf RE, Parker SC. The utility of pudendal nerve terminal motor latencies in idiopathic incontinence. Dis Colon Rectum 2006;49:852-7.

[80] Suilleabhain CB, Horgan AF, McEnroe L, Poon FW, Anderson JH, Finlay IG et al. The relationship of pudendal nerve terminal motor latency to squeeze pressure in patients with idiopathic fecal incontinence. Dis Colon Rectum 2001;44:666-71.

[81] Glickman S, Kamm MA. Bowel dysfunction in spinal-cord-injury patients. Lancet 1996;347:1651-3.

[82] Krogh K, Nielsen J, Djurhuus JC, Mosdal C, Sabroe S, Laurberg S. Colorectal function in patients with spinal cord lesions. Dis Colon Rectum 1997;40:1233-9.

[83] Lynch AC, Wong C, Anthony A, Dobbs BR, Frizelle FA. Bowel dysfunction following spinal cord injury: a description of bowel function in a spinal cord-injured population and comparison with age and gender matched controls. Spinal Cord 2000;38:717-23.

[84] Liu CW, Huang CC, Chen CH, Yang YH, Chen TW, Huang MH. Prediction of severe neurogenic bowel dysfunction in persons with spinal cord injury. Spinal Cord 2010;48:554-9.

[85] Han TR, Kim JH, Kwon BS. Chronic gastrointestinal problems and bowel dysfunction in patients with spinal cord injury. Spinal Cord 1998;36:485-90.

[86] Kannisto M, Rintala R. Bowel function in adults who have sustained spinal cord injury in childhood. Paraplegia 1995;33:701-3.

[87] Valles M, Vidal J, Clave P, Mearin F. Bowel dysfunction in patients with motor complete spinal cord injury: clinical, neurological, and pathophysiological associations. Am J Gastroenterol 2006;101:2290-9.

[88] Clinton N, Prott G, Rutkowski S, Li YM, Hansen R, Kellow J et al. Gastrointestinal symptoms in spinal cord injury: Relationships with level of injury and psychologic factors. Diseases of the Colon & Rectum 2005;48:1562-8.

[89] Stone JM, Nino-Murcia M, Wolfe VA, Perkash I. Chronic gastrointestinal problems in spinal cord injury patients: a prospective analysis. Am J Gastroenterol 1990;85:1114-9.

[90] Faaborg PM, Christensen P, Finnerup N, Laurberg S, Krogh K. The pattern of colorectal dysfunction changes with time since spinal cord injury. Spinal Cord 2008;46:234-8.

[91] Krogh K, Mosdal C, Gregersen H, Laurberg S. Rectal wall properties in patients with acute and chronic spinal cord lesions. Dis Colon Rectum 2002;45:641-9.

[92] Lynch AC, Anthony A, Dobbs BR, Frizelle FA. Anorectal physiology following spinal cord injury. Spinal Cord 2000;38:573-80.

[93] Sun WM, MacDonagh R, Forster D, Thomas DG, Smallwood R, Read NW. Anorectal function in patients with complete spinal transection before and after sacral posterior rhizotomy. Gastroenterology 1995;108:990-8.

[94] Lynch AC, Antony A, Dobbs BR, Frizelle FA. Bowel dysfunction following spinal cord injury. Spinal Cord 2001;39:193-20.

[95] Krogh K, Mosdal C, Laurberg S. Gastrointestinal and segmental colonic transit times in patients with acute and chronic spinal cord lesions. Spinal Cord 2000;38:615-21.

[96] Keshavarzian A, Barnes WE, Bruninga K, Nemchausky B, Mermall H, Bushnell D. Delayed colonic transit in spinal cord-injured patients measured by indium-111 Amberlite scintigraphy. Am J Gastroenterol 1995;90:1295-300.

[97] Nino-Murcia M, Stone JM, Chang PJ, Perkash I. Colonic transit in spinal cord-injured patients. Invest Radiol 1990;25:109-12.

[98] Krogh K, Olsen N, Christensen P, Madsen JL, Laurberg S. Colorectal transport during defecation in patients with lesions of the sacral spinal cord. Neurogastroenterol Motil 2003;15:25-31.

[99] Karlsson AK. Autonomic dysreflexia. Spinal Cord 1999;37:383-91.

[100] Parkman HP, Jones MP. Tests of gastric neuromuscular function. Gastroenterology 2009;136:1526-43.

[101] Rao S, Lele V. Scintigraphy of the small intestine: a simplified standard for study of transit with reference to normal values. Eur J Nucl Med Mol Imaging 2002;29:971; author reply 971-2.

[102] Argenyi EE, Soffer EE, Madsen MT, Berbaum KS, Walkner WO. Scintigraphic evaluation of small bowel transit in healthy subjects: inter- and intrasubject variability. Am J Gastroenterol 1995;90:938-42.

[103] Tougas G, Eaker EY, Abell TL, Abrahamsson H, Boivin M, Chen J et al. Assessment of gastric emptying using a low fat meal: establishment of international control values. Am J Gastroenterol 2000;95:1456-62.

[104] Abell TL, Camilleri M, Donohoe K, Hasler WL, Lin HC, Maurer AH et al. Consensus recommendations for gastric emptying scintigraphy: a joint report of the American Neurogastroenterology and Motility Society and the Society of Nuclear Medicine. Am J Gastroenterol 2008;103:753-6.

[105] Gregersen H, Orvar K, Christensen J. Biomechanical properties of duodenal wall and duodenal tone during phase I and phase II of the MMC. Am J Physiol 1992;263:G795-801.

[106] Camilleri M, Hasler WL, Parkman HP, Quigley EM, Soffer E. Measurement of gastrointestinal motility in the GI laboratory. Gastroenterology 1998;115:747-62.

[107] Verhagen MA, Samsom M, Jebbink RJ, Smout AJ. Clinical relevance of antroduodenal manometry. Eur J Gastroenterol Hepatol 1999;11:523-8.

[108] Malagelada JR, Stanghellini V. Manometric evaluation of functional upper gut symptoms. Gastroenterology 1985;88:1223-31.

[109] Rao SS, Kuo B, McCallum RW, Chey WD, DiBaise JK, Hasler WL et al. Investigation of colonic and whole-gut transit with wireless motility capsule and radiopaque markers in constipation. Clin Gastroenterol Hepatol 2009;7:537-44.

[110] Cassilly D, Kantor S, Knight LC, Maurer AH, Fisher RS, Semler J et al. Gastric emptying of a non-digestible solid: assessment with simultaneous SmartPill pH and pressure capsule, antroduodenal manometry, gastric emptying scintigraphy. Neurogastroenterol Motil 2008;20:311-9.

[111] Rao SS, Mysore K, Attaluri A, Valestin J. Diagnostic Utility of Wireless Motility Capsule in Gastrointestinal Dysmotility. J Clin Gastroenterol 2010.

[112] Kloetzer L, Chey WD, McCallum RW, Koch KL, Wo JM, Sitrin M et al. Motility of the antroduodenum in healthy and gastroparetics characterized by wireless motility capsule. Neurogastroenterol Motil 2010;22:527,33, e117.

[113] Gregersen H, Andersen MB. Impedance measuring system for quantification of cross-sectional area in the gastrointestinal tract. Med Biol Eng Comput 1991;29:108-10.

[114] Pescatori M, Anastasio G, Bottini C, Mentasti A. New grading and scoring for anal incontinence. Evaluation of 335 patients. Dis Colon Rectum 1992;35:482-7.

[115] Lunniss PJ, Kamm MA, Phillips RK. Factors affecting continence after surgery for anal fistula. Br J Surg 1994;81:1382-5.

[116] Rockwood TH, Church JM, Fleshman JW, Kane RL, Mavrantonis C, Thorson AG et al. Patient and surgeon ranking of the severity of symptoms associated with fecal incontinence: the fecal incontinence severity index. Dis Colon Rectum 1999;42:1525-32.

[117] Hull TL, Floruta C, Piedmonte M. Preliminary results of an outcome tool used for evaluation of surgical treatment for fecal incontinence. Dis Colon Rectum 2001;44:799-805.

[118] Bai Y, Chen H, Hao J, Huang Y, Wang W. Long-term outcome and quality of life after the Swenson procedure for Hirschsprung's disease. J Pediatr Surg 2002;37:639-42.

[119] Vaizey CJ, Carapeti E, Cahill JA, Kamm MA. Prospective comparison of faecal incontinence grading systems. Gut 1999;44:77-80.

[120] Jorge JM, Wexner SD. Etiology and management of fecal incontinence. Dis Colon Rectum 1993;36:77-9.

[121] Guyatt GH, Veldhuyzen Van Zanten SJ, Feeny DH, Patrick DL. Measuring quality of life in clinical trials: a taxonomy and review. CMAJ 1989;140:1441-8.

[122] Wood-Dauphinee S. Assessing quality of life in clinical research: from where have we come and where are we going? J Clin Epidemiol 1999;52:355-63.

[123] Soffer EE, Hull T. Fecal incontinence: a practical approach to evaluation and treatment. Am J Gastroenterol 2000;95:1873-80.

[124] Deutekom M, Terra MP, Dobben AC, Dijkgraaf MG, Felt-Bersma RJ, Stoker J et al. Selecting an outcome measure for evaluating treatment in fecal incontinence. Dis Color Rectum 2005;48:2294-301.

[125] Maeda Y, Pares D, Norton C, Vaizey CJ, Kamm MA. Does the St. Mark's incontinence score reflect patients' perceptions? A review of 390 patients. Dis Colon Rectum 2008;51:436-42.

[126] Rockwood TH, Church JM, Fleshman JW, Kane RL, Mavrantonis C, Thorson AG et al. Fecal Incontinence Quality of Life Scale: quality of life instrument for patients with fecal incontinence. Dis Colon Rectum 2000;43:9,16; discussion 16-7.

[127] Krogh K, Christensen P, Sabroe S, Laurberg S. Neurogenic bowel dysfunction score. Spinal Cord 2006;44:625-31.

[128] Bartram CI, Sultan AH. Anal endosonography in faecal incontinence. Gut 1995;37:4-6.

[129] Bartram C. Radiologic evaluation of anorectal disorders. Gastroenterol Clin North Am 2001;30:55-7.

[130] Sentovich SM, Blatchford GJ, Rivela LJ, Lin K, Thorson AG, Christensen MA. Diagnosing anal sphincter injury with transanal ultrasound and manometry. Dis Colon Rectum 1997;40:1430-4.

[131] Stoker J, Halligan S, Bartram CI. Pelvic floor imaging. Radiology 2001;218:621-4.

[132] Griffiths D. The pressure within a collapsed tube, with special reference to urethral pressure. Phys Med Biol 1985;30:951-63.

[133] Simpson RR, Kennedy ML, Nguyen MH, Dinning PG, Lubowski DZ. Anal manometry: a comparison of techniques. Dis Colon Rectum 2006;49:1033-8.

[134] Bharucha AE, Seide B, Fox JC, Zinsmeister AR. Day-to-day reproducibility of anorectal sensorimotor assessments in healthy subjects. Neurogastroenterol Motil 2004;16:241-50.

[135] Cali RL, Blatchford GJ, Perry RE, Pitsch RM, Thorson AG, Christensen MA. Normal variation in anorectal manometry. Dis Colon Rectum 1992;35:1161-4.

[136] Felt-Bersma RJ, Gort G, Meuwissen SG. Normal values in anal manometry and rectal sensation: a problem of range. Hepatogastroenterology 1991;38:444-9.

[137] Kiff ES, Swash M. Slowed conduction in the pudendal nerves in idiopathic (neurogenic) faecal incontinence. Br J Surg 1984;71:614-6.

[138] Tetzschner T, Sorensen M, Rasmussen OO, Lose G, Christiansen J. Reliability of pudendal nerve terminal motor latency. Int J Colorectal Dis 1997;12:280-4.

[139] Keighley MR, Henry MM, Bartolo DC, Mortensen NJ. Anorectal physiology measurement: report of a working party. Br J Surg 1989;76:356-7.

[140] Ryhammer AM, Laurberg S, Bek KM. Age and anorectal sensibility in normal women. Scand J Gastroenterol 1997;32:278-84.

[141] Metcalf AM, Phillips SF, Zinsmeister AR, MacCarty RL, Beart RW, Wolff BG. Simplified assessment of segmental colonic transit. Gastroenterology 1987;92:40-7.

[142] Abrahamsson H, Antov S, Bosaeus I. Gastrointestinal and colonic segmental transit time evaluated by a single abdominal x-ray in healthy subjects and constipated patients. Scand J Gastroenterol Suppl 1988;152:72-80.

[143] Sasaki Y, Hada R, Nakajima H, Munakata A. Difficulty in estimating localized bowel contraction by colonic manometry: a simultaneous recording of intraluminal pressure and luminal calibre. Neurogastroenterol Motil 1996;8:247-53.

[144] Toma TP, Zighelboim J, Phillips SF, Talley NJ. Methods for studying intestinal sensitivity and compliance: in vitro studies of balloons and a barostat. Neurogastroenterol Motil 1996;8:19-28.

[145] Madoff RD, Orrom WJ, Rothenberger DA, Goldberg SM. Rectal compliance: a critical reappraisal. Int J Colorectal Dis 1990;5:37-40.

[146] Zbar AP. Compliance and capacity of the normal human rectum--physical considerations and measurement pitfalls. Acta Chir Iugosl 2007;54:49-57.

[147] Krogh K, Ryhammer AM, Lundby L, Gregersen H, Laurberg TS. Comparison of methods used for measurement of rectal compliance. Dis Colon Rectum 2001;44:199-206.

[148] Gregersen H, Djurhuus JC. Impedance planimetry: a new approach to biomechanical intestinal wall properties. Dig Dis 1991;9:332-40.

[149] Petersen P, Gao C, Rossel P, Qvist P, Arendt-Nielsen L, Gregersen H et al. Sensory and biomechanical responses to distension of the normal human rectum and sigmoid colon. Digestion 2001;64:191-9.

[150] Drewes AM, Frokjaer JB, Larsen E, Reddy H, Arendt-Nielsen L, Gregersen H. Pain and mechanical properties of the rectum in patients with active ulcerative colitis. Inflamm Bowel Dis 2006;12:294-303.

[151] Andersen IS, Michelsen HB, Krogh K, Buntzen S, Laurberg S. Impedance Planimetric Description of Normal Rectoanal Motility in Humans. Dis Colon Rectum 2007.

[152] Krogh K. Colorectal function in patients with spinal cord lesions. 2000.

[153] Andersen IS, Michelsen HB, Krogh K, Buntzen S, Laurberg S. Impedance planimetric description of normal rectoanal motility in humans. Dis Colon Rectum 2007;50:1840-8.

[154] Andersen IS, Gregersen H, Buntzen S, Djurhuus JC, Laurberg S. New probe for the measurement of dynamic changes in the rectum. Neurogastroenterol Motil 2004;16:99-105.

[155] Malone PS, Ransley PG, Kiely EM. Preliminary report: the antegrade continence enema. Lancet 1990;336:1217-8.

[156] Baeten CG, Bailey HR, Bakka A, Belliveau P, Berg E, Buie WD et al. Safety and efficacy of dynamic graciloplasty for fecal incontinence: report of a prospective, multicenter trial. Dynamic Graciloplasty Therapy Study Group. Dis Colon Rectum 2000;43:743-51.

[157] Lehur PA, Glemain P, Bruley des Varannes S, Buzelin JM, Leborgne J. Outcome of patients with an implanted artificial anal sphincter for severe faecal incontinence. A single institution report. Int J Colorectal Dis 1998;13:88-92.

[158] Norton C, Cody JD, Hosker G. Biofeedback and/or sphincter exercises for the treatment of faecal incontinence in adults. Cochrane database of systematic reviews (Online) 2006;3.

[159] Deutekom M, Dobben A. Plugs for containing faecal incontinence. Cochrane Database Syst Rev 2005;(3):CD005086.

[160] Maeda Y, Laurberg S, Norton C. Perianal injectable bulking agents as treatment for faecal incontinence in adults. Cochrane Database Syst Rev 2010;5:CD007959.

[161] Hosker G, Cody JD, Norton CC. Electrical stimulation for faecal incontinence in adults. Cochrane Database Syst Rev 2007;(3):CD001310.

[162] Brown SR, Nelson RL. Surgery for faecal incontinence in adults. Cochrane Database Syst Rev 2007;(2):CD001757.

[163] Mowatt G, Glazener C, Jarrett M. Sacral nerve stimulation for faecal incontinence and constipation in adults. Cochrane Database Syst Rev 2007;-:CD004464.

[164] Lawrence JM, Lukacz ES, Nager CW, Hsu JW, Luber KM. Prevalence and co-occurrence of pelvic floor disorders in community-dwelling women. Obstet Gynecol 2008;111:678-85.

[165] Erekson EA, Sung VW, Myers DL. Effect of body mass index on the risk of anal incontinence and defecatory dysfunction in women. Am J Obstet Gynecol 2008;198:596.e1,596.e4.

[166] Lauti M, Scott D, Thompson-Fawcett MW. Fibre supplementation in addition to loperamide for faecal incontinence in adults: a randomized trial. Colorectal Dis 2008;10:553-62.

[167] Bliss DZ, Jung HJ, Savik K, Lowry A, LeMoine M, Jensen L et al. Supplementation with dietary fiber improves fecal incontinence. Nurs Res 2001;50:203-1.

[168] Chatoor DR, Taylor SJ, Cohen CR, Emmanuel AV. Faecal incontinence. Br J Surg 2007;94:134-4.

[169] Papachrysostomou MC, Smith AN, Stevenson AJM. Does electrical stimulation of the pelvic floor alter the proctographic parameters in faecal incontinence? Eur J Gastroenterol Hepatol 1994;6:139-44.

[170] Wald A. Biofeedback therapy for fecal incontinence. Ann Intern Med 1981;95:146-9.

[171] Buser WD, Miner PB,Jr. Delayed rectal sensation with fecal incontinence. Successful treatment using anorectal manometry. Gastroenterology 1986;91:1186-91.

[172] Norton C, Kamm MA. Anal sphincter biofeedback and pelvic floor exercises for faecal incontinence in adults--a systematic review. Aliment Pharmacol Ther 2001;15:1147-54.

[173] Whitehead WE, Wald A, Norton NJ. Treatment options for fecal incontinence. Dis Colon Rectum 2001;44:131,42; discussion 142-4.

[174] Norton C, Chelvanayagam S, Wilson-Barnett J, Redfern S, Kamm MA. Randomized controlled trial of biofeedback for fecal incontinence. Gastroenterology 2003;125:1320-9.

[175] Christensen P, Olsen N, Krogh K, Bacher T, Laurberg S. Scintigraphic assessment of retrograde colonic washout in fecal incontinence and constipation. Dis Colon Rectum 2003;46:68-76.

[176] Christensen P, Krogh K, Buntzen S, Payandeh F, Laurberg S. Long-term outcome and safety of transanal irrigation for constipation and fecal incontinence. Dis Colon Rectum 2009;52:286-92.

[177] Christensen P, Bazzocchi G, Coggrave M, Abel R, Hultling C, Krogh K et al. A randomized, controlled trial of transanal irrigation versus conservative

bowel management in spinal cord-injured patients. Gastroenterology 2006;131:738-47.

[178] Faaborg PM, Christensen P, Kvitsau B, Buntzen S, Laurberg S, Krogh K. Long-term outcome and safety of transanal colonic irrigation for neurogenic bowel dysfunction. Spinal Cord 2009;47:545-9.

[179] Christensen P, Buntzen S, Krogh K, Laurberg S. Ileal neoappendicostomy for antegrade colonic irrigation. Br J Surg 2001;88:1637-8.

[180] Fonkalsrud EW, Dunn JCY, Kawaguchi AI. Simplified technique for antegrade continence enemas for fecal retention and incontinence. J Am Coll Surg 1998;187:457-60.

[181] Gerharz EW, Vik V, Webb G, Woodhouse CR. The in situ appendix in the Malone antegrade continence enema procedure for faecal incontinence. Br J Urol 1997;79:985-6.

[182] Herndon CD, Cain MP, Casale AJ, Rink RC. The colon flap/extension Malone antegrade continence enema: an alternative to the Monti-Malone antegrade continence enema. J Urol 2005;174:299-302.

[183] Kurzrock EA, Karpman E, Stone AR. Colonic tubes for the antegrade continence enema: comparison of surgical technique. J Urol 2004;172:700-2.

[184] Monti PR, Lara RC, Dutra MA, de Carvalho JR. New techniques for construction of efferent conduits based on the Mitrofanoff principle. Urology 1997;49:112-5.

[185] Tackett LD, Minevich E, Benedict JF, Wacksman J, Sheldon CA. Appendiceal versus ileal segment for antegrade continence enema. J Urol 2002;167:683-6.

[186] Weiser AC, Stock JA, Hanna MK. Modified cecal flap neoappendix for the Malone antegrade continence enema procedure: a novel technique. J Urol 2003;169:2321-4.

[187] Worsoe J, Christensen P, Krogh K, Buntzen S, Laurberg S. Long-term results of antegrade colonic enema in adult patients: assessment of functional results. Dis Colon Rectum 2008;51:1523-8.

[188] Furlan JC, Urbach DR, Fehlings MG. Optimal treatment for severe neurogenic bowel dysfunction after chronic spinal cord injury: a decision analysis. Br J Surg 2007;94:1139-50.

[189] Kotanagi H, Koyama K, Sato Y, Takahashi K. Appendicostomy irrigation for facilitating colonic evacuation in colostomy patients. Preliminary report. Dis Colon Rectum 1998;41:1050,2; discussion 1052-3.

[190] Parks AG. Royal Society of Medicine, Section of Proctology; Meeting 27 November 1974. President's Address. Anorectal incontinence. Proc R Soc Med 1975;68:681-90.

[191] Osterberg A, Edebol Eeg-Olofsson K, Graf W. Results of surgical treatment for faecal incontinence. Br J Surg 2000;87:1546-52.

[192] Abbas SM, Bissett IP, Neill ME, Parry BR. Long-term outcome of postanal repair in the treatment of faecal incontinence. ANZ J Surg 2005;75:783-6.

[193] Mackey P, Mackey L, Kennedy ML, King DW, Newstead GL, Douglas PR et al. Postanal repair--do the long-term results justify the procedure? Colorectal Dis 2010;12:367-72.

[194] Baeten CG, Geerdes BP, Adang EM, Heineman E, Konsten J, Engel GL et al. Anal dynamic graciloplasty in the treatment of intractable fecal incontinence. N Engl J Med 1995;332:1600-5.

[195] Rongen MJ, Uludag O, El Naggar K, Geerdes BP, Konsten J, Baeten CG. Long-term follow-up of dynamic graciloplasty for fecal incontinence. Dis Colon Rectum 2003;46:716-21.

[196] O'Brien PE, Dixon JB, Skinner S, Laurie C, Khera A, Fonda D. A prospective, randomized, controlled clinical trial of placement of the artificial bowel sphincter (Acticon Neosphincter) for the control of fecal incontinence. Dis Colon Rectum 2004;47:1852-60.

[197] Bortolotti M, Ugolini G, Grandis A, Montroni I, Mazzero G. A novel magnetic device to prevent fecal incontinence (preliminary study). Int J Colorectal Dis 2008;23:499-501.

[198] Norton C, Burch J, Kamm MA. Patients' views of a colostomy for fecal incontinence. Dis Colon Rectum 2005;48:1062-9.

[199] Colquhoun P, Kaiser R,Jr, Efron J, Weiss EG, Nogueras JJ, Vernava AM,3rd et al. Is the quality of life better in patients with colostomy than patients with fecal incontinence? World J Surg 2006;30:1925-8.

[200] Matzel KE, Kamm MA, Stosser M, Baeten CGMI, Christiansen J, Madoff R et al. Sacral spinal nerve stimulation for faecal incontinence: Multicentre study. Lancet 2004;363:1270-6.

[201] Rasmussen OO, Buntzen S, Sorensen M, Laurberg S, Christiansen J. Sacral nerve stimulation in fecal incontinence. Dis Colon Rectum 2004;47:1158-62.

[202] Jarrett MED, Varma JS, Duthie GS, Nicholls RJ, Kamm MA. Sacral nerve stimulation for faecal incontinence in the UK. Br J Surg 2004;91:755-61.

[203] Leroi AM, Parc Y, Lehur PA, Mion F, Barth X, Rullier E et al. Efficacy of sacral nerve stimulation for fecal incontinence: Results of a multicenter double-blind crossover study. Ann Surg 2005;242:662-9.

[204] Holzer B, Rosen HR, Novi G, Ausch C, Hölbling N, Schiessel R. Sacral nerve stimulation for neurogenic faecal incontinence. Br J Surg 2007;94:749-53.

[205] Melenhorst J, Koch SM, Uludag Ö, Van Gemert WG, Baeten CG. Sacral neuromodulation in patients with faecal incontinence: Results of the first 100 permanent implantations. Colorectal Disease 2007;9:725-30.

[206] Hollingshead JR, Dudding TC, Vaizey CJ. Sacral nerve stimulation for faecal incontinence: results from a single centre over a 10 year period. Colorectal Dis 2010.

[207] Matzel KE, Lux P, Heuer S, Besendorfer M, Zhang W. Sacral nerve stimulation for faecal incontinence: long-term outcome. Colorectal Dis 2009;11:636-41.

[208] Michelsen HB, Thompson-Fawcett M, Lundby L, Krogh K, Laurberg S, Buntzen S. Six years of experience with sacral nerve stimulation for fecal incontinence. Dis Colon Rectum 2010;53:414-21.

[209] Maeda Y, Lundby L, Buntzen S, Laurberg S. Suboptimal outcome following sacral nerve stimulation for faecal incontinence. Br J Surg 2011;98:140-7.

[210] Tjandra JJ, Lim JF, Matzel K. Sacral nerve stimulation: An emerging treatment for faecal incontinence. ANZ J Surg 2004;74:1098-106.

[211] Ratto C, Grillo E, Parello A, Petrolino M, Costamagna G, Doglietto GB. Sacral neuromodulation in treatment of fecal incontinence following anterior resection and chemoradiation for rectal cancer. Dis Colon Rectum 2005;48:1027-36.

[212] Jarrett ME, Matzel KE, Stosser M, Christiansen J, Rosen H, Kamm MA. Sacral nerve stimulation for faecal incontinence following a rectosigmoid resection for colorectal cancer. Int J Colorectal Dis 2005;20:446-51.

[213] Jarrett ME, Matzel KE, Stosser M, Baeten CG, Kamm MA. Sacral nerve stimulation for fecal incontinence following surgery for rectal prolapse repair: a multicenter study. Dis Colon Rectum 2005;48:1243-8.

[214] Jarrett ME, Matzel KE, Christiansen J, Baeten CG, Rosen H, Bittorf B et al. Sacral nerve stimulation for faecal incontinence in patients with previous partial spinal injury including disc prolapse. Br J Surg 2005;92:734-9.

[215] [215] Jarrett ME. Neuromodulation for constipation and fecal incontinence. Urol Clin North Am 2005;32:79-87.

[216] Vitton V, Gigout J, Grimaud JC, Bouvier M, Desjeux A, Orsoni P. Sacral nerve stimulation can improve continence in patients with Crohn's disease with internal and external anal sphincter disruption. Dis Colon Rectum 2008;51:924-7.

[217] Lundby L, Krogh K, Buntzen S, Laurberg S. Temporary sacral nerve stimulation for treatment of irritable bowel syndrome: a pilot study. Dis Colon Rectum 2008;51:1074-8.

[218] Plevnik S, Vodušek DB, Vrtačnik P, Janež J. Optimization of pulse duration for electrical stimulation in treatment of urinary incontinence. World Journal of Urology 1986;4:22-3.

[219] McCreery DB, Agnew WF, Yuen TG, Bullara LA. Relationship between stimulus amplitude, stimulus frequency and neural damage during electrical stimulation of sciatic nerve of cat. Med Biol Eng Comput 1995;33:426-9.

[220] Gorman PH, Mortimer JT. The effect of stimulus parameters on the recruitment characteristics of direct nerve stimulation. IEEE Trans Biomed Eng 1983;30:407-14.

[221] Cooperberg MR, Stoller ML. Percutaneous neuromodulation. Urol Clin North Am 2005;32:71-8.

[222] Salmons S, Vrbova G. The influence of activity on some contractile characteristics of mammalian fast and slow muscles. J Physiol 1969;201:535-49.

[223] Hudlicka O, Dodd L, Renkin EM, Gray SD. Early changes in fiber profile and capillary density in long-term stimulated muscles. Am J Physiol 1982;243:H528-35.

[224] Hosker G, Cody JD, Norton CC. Electrical stimulation for faecal incontinence in adults. Cochrane Database Syst Rev 2007;(3):CD001310.

[225] McGuire EJ, Zhang SC, Horwinski ER, Lytton B. Treatment of motor and sensory detrusor instability by electrical stimulation. J Urol 1983;129:78-9.

[226] Klingler HC, Pycha A, Schmidbauer J, Marberger M. Use of peripheral neuromodulation of the S3 region for treatment of detrusor overactivity: a urodynamic-based study. Urology 2000;56:766-71.

[227] van Balken MR, Vandoninck V, Gisolf KW, Vergunst H, Kiemeney LA, Debruyne FM et al. Posterior tibial nerve stimulation as neuromodulative treatment of lower urinary tract dysfunction. J Urol 2001;166:914-8.

[228] van Balken MR, Vandoninck V, Messelink BJ, Vergunst H, Heesakkers JP, Debruyne FM et al. Percutaneous tibial nerve stimulation as neuromodulative treatment of chronic pelvic pain. Eur Urol 2003;43:158,63; discussion 163.

[229] Vandoninck V, van Balken MR, Finazzi Agro E, Petta F, Micali F, Heesakkers JP et al. Percutaneous tibial nerve stimulation in the treatment of overactive bladder: urodynamic data. Neurourol Urodyn 2003;22:227-32.

[230] Vandoninck V, van Balken MR, Finazzi Agro E, Petta F, Micali F, Heesakkers JP et al. Posterior tibial nerve stimulation in the treatment of idiopathic nonobstructive voiding dysfunction. Urology 2003;61:567-72.

[231] Nakamura M, Sakurai T, Tsujimoto Y, Tada Y. Transcutaneous electrical stimulation for the control of frequency and urge incontinence. Hinyokika Kiyo 1983;29:1053-9.

[232] Previnaire JG, Soler JM, Perrigot M, Boileau G, Delahaye H, Schumacker P et al. Short-term effect of pudendal nerve electrical stimulation on detrusor hyperreflexia in spinal cord injury patients: importance of current strength. Paraplegia 1996;34:95-9.

[233] Kirkham AP, Shah NC, Knight SL, Shah PJ, Craggs MD. The acute effects of continuous and conditional neuromodulation on the bladder in spinal cord injury. Spinal Cord 2001;39:420-8.

[234] Dalmose AL, Rijkhoff NJ, Kirkeby HJ, Nohr M, Sinkjaer T, Djurhuus JC. Conditional stimulation of the dorsal penile/clitoral nerve may increase cystometric capacity in patients with spinal cord injury. Neurourol Urodyn 2003;22:130-7.

[235] Hansen J, Media S, Nohr M, Biering-Sorensen F, Sinkjaer T, Rijkhoff NJ. Treatment of neurogenic detrusor overactivity in spinal cord injured patients by conditional electrical stimulation. J Urol 2005;173:2035-9.

[236] Fjorback MV, Rijkhoff N, Petersen T, Nohr M, Sinkjaer T. Event driven electrical stimulation of the dorsal penile/clitoral nerve for management of neurogenic detrusor overactivity in multiple sclerosis. Neurourol Urodyn 2006;25:349-55.

[237] McFarlane JP, Foley SJ, de Winter P, Shah PJ, Craggs MD. Acute suppression of idiopathic detrusor instability with magnetic stimulation of the sacral nerve roots. Br J Urol 1997;80:734-41.

[238] Craggs MD, Balasubramaniam AV, Chung EA, Emmanuel AV. Aberrant reflexes and function of the pelvic organs following spinal cord injury in man. Auton Neurosci 2006;126-127:355-70.

[239] Binnie NR, Kawimbe BM, Papachryosostomou M, Smith AN. Use of the pudendo-anal reflex in the treatment of neurogenic faecal incontinence. Gut 1990;31:1051-5.

[240] Frizelle FA, Gearry RB, Johnston M, Barclay ML, Dobbs BR, Wise C et al. Penile and clitoral stimulation for faecal incontinence: External application of a bipolar electrode for patients with faecal incontinence. Colorectal Disease 2004;6:54-7.

[241] Shafik A, Ahmed I, El-Sibai O, Mostafa RM. Percutaneous peripheral neuromodulation in the treatment of fecal incontinence. Eur Surg Res 2003;35:103-7.

[242] Queralto M, Portier G, Cabarrot PH, Bonnaud G, Chotard JP, Nadrigny M et al. Preliminary results of peripheral transcutaneous neuromodulation in the treatment of idiopathic fecal incontinence. Int J Colorectal Dis 2006;21:670-2.

[243] Mentes BB, Yuksel O, Aydin A, Tezcaner T, Leventoglu A, Aytac B. Posterior tibial nerve stimulation for faecal incontinence after partial spinal injury: Preliminary report. Tech Coloproctol 2007;11:115-9.

[244] de la Portilla F, Rada R, Vega J, Gonzalez CA, Cisneros N, Maldonado VH. Evaluation of the use of posterior tibial nerve stimulation for the treatment of fecal incontinence: preliminary results of a prospective study. Dis Colon Rectum 2009;52:1427-33.

[245] Vitton V, Damon H, Roman S, Nancey S, Flourié B, Mion F. Transcutaneous posterior tibial nerve stimulation for fecal incontinence in inflammatory bowel disease patients: A therapeutic option? Inflamm Bowel Dis 2009;15:402-5.

[246] Boyle DJ, Prosser K, Allison ME, Williams NS, Chan CL. Percutaneous tibial nerve stimulation for the treatment of urge fecal incontinence. Dis Colon Rectum 2010;53:432-7.

[247] Findlay JM, Yeung JM, Robinson R, Greaves H, Maxwell-Armstrong C. Peripheral neuromodulation via posterior tibial nerve stimulation - a potential treatment for faecal incontinence? Ann R Coll Surg Engl 2010;92:385-90.

[248] Govaert B, Pares D, Delgado-Aros S, La Torre F, van Gemert W, Baeten C. A Prospective Multicenter Study to investigate Percutaneous Tibial Nerve Stimulation for the Treatment of Faecal Incontinence. Colorectal Dis 2009;[Epub ahead of print].

[249] Eleouet M, Siproudhis L, Guillou N, Le Couedic J, Bouguen G, Bretagne JF. Chronic posterior tibial nerve transcutaneous electrical nerve stimulation (TENS) to treat fecal incontinence (FI). Int J Colorectal Dis 2010;25:1127-32.

[250] Stoller ML. Afferent nerve stimulation for pelvic floor dysfunction. Eur Urol 1999;35:16.

[251] Fynes MM, Marshall K, Cassidy M, Behan M, Walsh D, O'Connell PR et al. A prospective, randomized study comparing the effect of augmented biofeedback with sensory biofeedback alone on fecal incontinence after obstetric trauma. Dis Colon Rectum 1999;42:753,8; discussion 758-61.

[252] Surh S, Kienle P, Stern J, Herfarth C. Passive electrostimulation therapy of the anal sphincter is inferior to active biofeedback training. Langenbecks Arch Chir Suppl Kongressbd 1998;115:976-8.

[253] Kienle P, Weitz J, Koch M, Benner A, Herfarth C, Schmidt J. Biofeedback versus electrostimulation in treatment of anal sphincter insufficiency. Dig Dis Sci 2003;48:1607-13.

[254] Scheuer M, Kuijpers HC, Bleijenberg G. Effect of electrostimulation on sphincter function in neurogenic fecal continence. Dis Colon Rectum 1994;37:590,3; discussion 593-4.

[255] Österberg A, Graf W, Eeg-Olofsson K, Hålldén M, Påhlman L. Is electrostimulation of the pelvic floor an effective treatment for neurogenic faecal incontinence? Scand J Gastroenterol 1999;34:319-24.

[256] Healy CF, Brannigan AE, Connolly EM, Eng M, O'Sullivan MJ, McNamara DA et al. The effects of low-frequency endo-anal electrical stimulation on faecal incontinence: A prospective study. Int J Colorectal Dis 2006;21:802-6.

[257] Mahony RT, Malone PA, Nalty J, Behan M, O'Connell PR, O'Herlihy C. Randomized clinical trial of intra-anal electromyographic biofeedback physiotherapy with intra-anal electromyographic biofeedback augmented with electrical stimulation of the anal sphincter in the early treatment of postpartum fecal incontinence. Obstet Gynecol 2004;191:885-90.

[258] Norton C, Gibbs A, Kamm MA. Randomized, controlled trial of anal electrical stimulation for fecal incontinence. Dis Colon Rectum 2006;49:190-6.

[259] Österberg A, Edebol Eeg-Olofsson K, Hålldén M, Graf W. Randomized clinical trial comparing conservative and surgical treatment of neurogenic faecal incontinence. Br J Surg 2004;91:1131-7.

[260] Haskell B, Rovner H. Electromyography in the management of the incompetent anal sphincter. Dis Colon Rectum 1967;10:81-4.

[261] Nie Y, Chen JDZ. Effects and mechanisms of anal electrical stimulation on anorectal compliance and tone in dogs. Dis Colon Rectum 2006;49:1414-21.

[262] Nie YQ, Pasricha JP, Chen JDZ. Anal electrical stimulation with long pulses increases anal sphincter pressure in conscious dogs. Diseases of the Colon & Rectum 2006;49:383-91.

[263] Song G-, Zhu H, Chen JDZ. Effects and mechanisms of vaginal electrical stimulation on rectal tone and anal sphincter pressure. Dis Colon Rectum 2007;50:2104-11.

[264] Gallas S, Gourcerol G, Ducrotté P, Mosni G, Menard J-, Michot F et al. Does magnetic stimulation of sacral nerve roots modify colonic motility?

Results of a randomized double-blind sham-controlled study. Neurogastroenterology and Motility 2009;21:411-9.

Chapter 2

Electrical stimulation of the dorsal clitoral nerve reduces incontinence episodes in idiopathic faecal incontinent patients: A pilot study

J. Worsøe[#], L. Fynne[+], S. Laurberg[*], K. Krogh[+], N.J.M. Rijkhoff[#]*

[#]*Center for Sensory-Motor Interaction (SMI), Department of Health Science and Technology, Aalborg University, Denmark*
[*]*Department of Surgery P, Aarhus University Hospital, Aarhus, Denmark*
[+]*Neurogastroenterology Unit, Department of Hepatology and Gastroenterology V, Aarhus University Hospital, Aarhus, Denmark*

2.1 INTRODUCTION

Faecal incontinence (FI) results from various aetiologies with an estimated prevalence of 4.3% in both genders and across all age groups [1]. Involuntary excretion and leaking are common occurrences for those affected. The symptoms are often socially unacceptable, and if the problems are not addressed, they can lead to social withdrawal and isolation. FI or subsequent consequences represents a considerable economic burden on society [2]. Current treatment algorithms usually apply a stepwise strategy of techniques ranging from conservative treatment and irrigation, to reconstructive sphincter surgery or ultimately a stoma [3]. Electrical stimulation is also used in the treatment of FI. Both sacral nerve stimulation (Interstim Therapy) and posterior tibial nerve stimulation (Urgent PC) have been used in thousands of faecal incontinent patients with good results [4-7].

In this study, the effect of dorsal genital nerve stimulation on FI was investigated. DGN stimulation has been shown to inhibit the bladder and reduce overactive bladder symptoms [8,9]. Because of the common innervation of the bladder and rectum, similar effects were anticipated for the rectum. Two studies have described a positive effect of DGN stimulation on FI. In these studies a stimulation frequency of 1-2 Hz was used which resulted in improved sphincter function [10,11].

The aim of this study was to assess the effect of transcutaneous electrical DGN stimulation twice daily for three weeks, in patients suffering from idiopathic FI.

2.2 METHODS

2.2.1   Patients

This study was a prospective, descriptive study. Ten female patients, (median age: 60 years; range 34 to 68 years), with idiopathic FI, including both urgency FI and passive FI, who had been referred to a tertiary centre participated in this study. Patients with severe sphincter lesions, previous pelvic or anorectal surgery, organic anorectal disease, systemic diseases affecting bowel function or neurological disorders were excluded. The assessment at baseline consisted of a medical history, a three-week bowel habit diary, anal physiology testing, endoanal ultrasonography and radiographically determined colonic transit time. This study was conducted in accordance with the Helsinki Declaration II, and was approved by the Scientific Ethical Committee of Region Midtjylland (M-20070219). All patients gave their fully informed written consent.

2.2.2   Stimulation

Stimulation was performed using a battery powered handheld stimulator (Itouch Plus, TensCare, Epsom, UK). Monophasic square constant current pulses with a pulse duration of 200 µs at a pulse rate of 20 Hz were used. The patients were instructed regarding the operation of the stimulator and initially used it under supervision. One electrode (dimensions: 20 x 10 mm, Neuroline 700, Ambu, Ballerup, Denmark) was placed on the clitoris as a cathode, and a second electrode (diameter: 32 mm, PALS Platinium, Axelgaard, Lystrup, Denmark) was placed 2-3 cm lateral to the right labia major (Fig. 1). Precautions were taken to ensure good contact between skin and the electrodes, including removal of hair. The Neuroline 700 electrode was used a single time, and the PALS electrodes were used twice. Furthermore, the patients were instructed to place the electrodes correctly and to clean the skin before application of the electrodes. During the training session, the clitoral-anal reflex threshold was identified when possible. The patients were encouraged to set the amplitude as high as tolerable each time when stimulating at home.

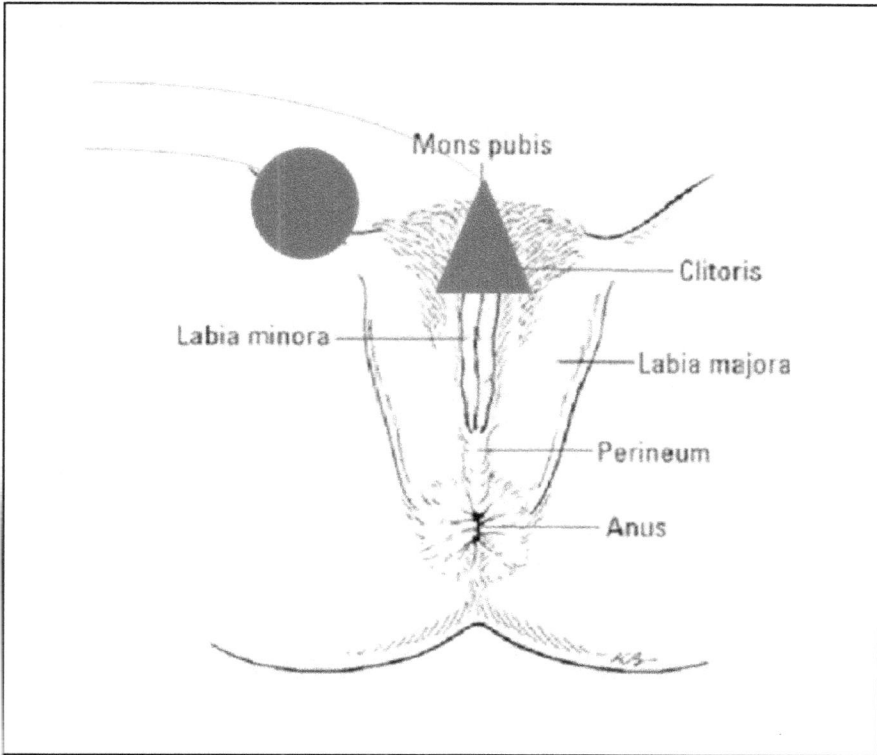

Figure 2.1: Application of the electrodes (schematic drawing). One electrode is placed on the clitoris as a cathode, and a second electrode is placed 2-3 cm lateral to the right labia major.

2.2.3   Protocol

Transcutaneous stimulation was performed twice daily for 15 minutes in each session for a period of three weeks. The patients recorded the stimulation amplitude after each session. Symptoms were assessed for three weeks before stimulation, during the three weeks of stimulation, and for three weeks after the final stimulation. The patients kept a bowel habit diary over the course of the study to collect information about the number of bowel movements, the number of urgency episodes, and the number of FI episodes. In each three week period, the patients were contacted by phone one time to ensure compliance using the bowel habit diary and with the stimulation procedure. After each period, the severity of FI symptoms was assessed using the Wexner and St. Mark's FI severity scores, and a self-reported FI severity visual analogue scale (VAS, 0-100, where 0 is the best and 100 is the worst) [12,13]. Disease-specific quality of life was evaluated using the Faecal incontinence quality of life score (FIQL; four scales: lifestyle, coping/behavior, depression/self-Perception, and embarrassment) [14]. Anal manometry (resting pressure and maximal voluntary squeeze pressure) and rectal distension parameters were also obtained at the end of each period.

2.2.4   Analysis

Wilcoxon's matched pairs test was used to compare the baseline results with the results obtained during and three weeks after stimulation. A p-value below 0.05 was considered statistically significant.

2.3 RESULTS

Nine patients completed the protocol. One patient withdrew for personal reasons. All patients exhibited a normal colonic transit time (<5 markers present after six days). According to transanal ultrasonography, seven had intact sphincters and two had minor defects. Left and right pudendal nerve terminal motor latency and anal sensation were within normal range. All of the patients found it easy to position the electrodes and control the stimulator. They also found it easy to find time for stimulation in their daily routine. No adverse events were reported. The median clitoro-anal reflex threshold was 9 mA (range: 8-25 mA). The median stimulation amplitude was 27 mA (range: 9-52 mA). Most patients increased the amplitude slightly during the stimulation period (median 1 mA; range 0.5-17 mA)), which indicated some adaption to the stimulation.   In four subjects, the stimulation amplitude was greater than two times the clitoral-anal reflex threshold (ID1, ID2, ID3, and ID7). Results are summarised in Tables 2.1 and 2.2.
   During the stimulation period, the number of bowel movements was lower in eight patients. This decrease was maintained for three weeks after stimulation in all eight patients. The median number of bowel movements decreased (p=0.038) from 45 at baseline to 40 during stimulation and to 39 at three weeks after stimulation (ns). The percentage of bowel movements with urgency decreased from 49% (range: 0-100) to 38% (range: 0-95) during stimulation (ns), and 23% (range: 0-100) for the three weeks after stimulation (p=0.018). The number of urgency episodes was lower in six patients during stimulation. The median number of urgency episodes decreased from 22 at baseline to 11 during stimulation (ns). One patient reported zero urgency episodes, and two patients experienced more urgency episodes during stimulation. Three weeks after stimulation, the number of urgency episodes decreased in eight patients and remained unchanged in one patient as compared to the baseline. The median number of urgency episodes during the three-week period after the final stimulation was 11 (p=0.025). Eight patients reported incontinence at baseline. The number of incontinence episodes was lower during stimulation in seven of the eight patients, and was maintained three weeks after stimulation in all the patients. The median number of incontinence episodes decreased from 13 at baseline to 2 during stimulation (P=0.025), and 5 for the three weeks after stimulation (P=0.017).

Patient ID	Stimulation amplitude (mA)	Number of bowel movements			Number of urgency episodes			Number of Incontinence episodes		
		Baseline	Stim	Post-stim	Baseline	Stim	Post-stim	Baseline	Stim	Post-stim
1	52	45	23	30	20	2	7	15	1	2
2	31	19	16	17	0	0	0	10	0	5
3	37	210	133	84	210	126	84	231	220	84
4	9	47	44	40	33	23	16	25	0	6
5	9	54	53	61	23	6	10	4	0	0
6	27	118	100	89	86	60	60	31	8	5
7	30	37	29	29	18	11	11	13	4	6
8	26	45	40	30	3	8	2	2	2	2
9	11	26	32	30	22	30	7	0	7	2
Median	27	45	*40	39	22	11	*11	13	*2	*5

Table 2.1. Data from bowel habit diaries recorded three weeks prior to stimulation (Baseline), three weeks during stimulation (Stim), and three weeks after the final stimulation (Post-stim). * p<0.05 compared to baseline

	Baseline	Stimulation	Post-stimulation
Wexner fecal incontinence score	14 (8-20)	*11 (4-16)	*11 (3-19)
St. Marks fecal incontinence score	16 (12-24)	*14 (4-18)	*13 (7-21)
VAS-score (1-100)	69 (42-100)	40 (7-100)	72 (30-100)
FIQL, lifestyle	2.89 (1.40-4.00)	3.10 (1.00-4.00)	3.20 (1.70-4.00)
FIQL, coping/ behavior	1.71 (1.00-2.88)	1.78 (1.00-3.56)	1.7 (1.00-3.63)
FIQL, depression/ self perception	2.88 (1.60-3.60)	2.56 (1.33-3.60)	2.68 (1.47-3.60)
FIQL, embarrassment	2.17 (1.67-4.00)	2.33 (1.00-3.67)	3.00 (1.33-5-00)
Anal resting pressure (cm H_2O)	62.5 (14-131)	46 (23-130)	48 (31-147)
Anal squeeze pressure (cm H_2O)	75.5 (12-146)	87 (32-177)	106 (25-156)
First sensation (ml)	65 (29-138)	58 (28-127)	70 (47-160)
Desire to defecate (ml)	90 (50-220)	104 (57-216)	95 (63-179)
Maximum tolerable rectal volume (ml)	170 (72-349)	155 (126-339)	172 (100-215)

Table 2.2: Wexner and St. Mark's scores, severity of FI VAS score (1-100), Faecal incontinence quality of life score (FIQL), anal resting and squeeze pressure, and rectal volume tolerability (first sensation, desire to defecate, and maximal tolerable volume) prior to stimulation, immediately after stimulation, and three weeks after stimulation. * p<0.05 compared to baseline.

Both the Wexner (p=0.027) and the St. Mark's scores (p=0.035) were significantly lower immediately after stimulation. Improvements in both the Wexner score (p=0.048), and the St. Mark's score (p=0.049) were maintained for three weeks after stimulation. Compared to the baseline, the Wexner score improved in seven patients during stimulation and in eight patients three weeks after stimulation. Compared to the baseline, the St. Mark's score was improved in six patients during stimulation and in seven patients three weeks after stimulation.

The patients' subjective assessment of FI severity using the VAS score was not improved with stimulation. FIQL scores for the four domains, lifestyle, coping/behaviour, depression/self-perception, and embarrassment, did not change during stimulation. Anal resting pressure, anal squeeze pressure, and rectal sensitivity assessed immediately after three weeks of stimulation and three weeks after the final stimulation were unchanged when compared to the baseline values.

Patients were contacted by phone one to six months after the end of the study. At that time, all of the patients reported that their symptoms had recurred after they had stopped using the stimulator, and all of the patients still required additional treatment for their FI symptoms.

2.4 DISCUSSION

This study shows that electrical stimulation of the clitoral nerve improves symptoms in patients suffering from FI. The effect was maintained for three weeks after the final stimulation, but a longer follow-up survey indicated that symptoms eventually reappeared. There was a significant reduction in the number of incontinence episodes, which was reflected in the Wexner and St. Mark's FI severity scores. Furthermore, during stimulation, there was a significant decrease in the overall number of bowel movements and the percentage of bowel movements with urgency. The improvement was still present three weeks after the final stimulation. Urgency is more frequent with loose stools, and it is conceivable that the decrease in the number of urgency episodes have been due to a change in stool consistency. However, this correlation could not be confirmed by the bowel habit diaries or the FI severity scores. The patients' subjective responses did not correspond with their diaries. The patients reported worsened symptoms within one week after the last stimulation, but an analysis of the diaries with respect to the number of events per week could not be used to estimate the duration of the effect of treatment due to a large range and limited number of events. In addition, significant improvement was not demonstrated in quality of life as assessed by the disease-specific FIQL or in patients' subjective assessment of FI severity using a VAS score. Finally, anal sphincter function and rectal sensitivity were unchanged after stimulation.

The severity of FI was evaluated with the Wexner score and the St. Mark's score, which are both widely used; the St. Mark's score has been demonstrated to correlate with patients' subjective perceptions [13,15,16]. The FIQL measures disease-specific impacts on four domains, lifestyle, coping/ behaviour, depression/

self-perception, and embarrassment [14]. Even though the patients experienced a reduction in symptoms, they did not become completely continent and therefore still felt restricted in daily activity and worried about incontinence episodes. From that perspective it seems unlikely that a stimulation period of only three weeks would be sufficient to significantly improve quality of life.

The lack of improvement of sphincter function and rectal volume tolerability during stimulation indicates that the decrease in symptoms was not due to a measurable strengthening of the sphincter apparatus or an improvement of rectal capacity. At both the initial assessment and at the assessment after stimulation, manometric values and values for rectal volume tolerability were within normal reference values for our laboratory [17]. However, resting pressure and squeeze pressure are not very sensitive to differences between continent and incontinent individuals, and data regarding the sensitivity and specificity of rectal balloon distension in distinguishing continent from incontinent patients are not available [18,19]. Other studies of electrical stimulation for the treatment of FI have evaluated sphincter function and rectal volume tolerability in successfully treated patients. However, no clear effects have been demonstrated and the results have been ambiguous, which indicates that improved sphincter function may not be necessary for improvement of FI symptoms [10,11,20-26].

The stimulation parameters in this study were chosen based on previous experiences with DGN stimulation to achieve inhibition of the bladder, in which a pulse width of 200 μs and a pulse rate of 20 Hz with an amplitude of at least twice the clitoral-anal reflex threshold was used to inhibit bladder contractions in patients with overactive bladder symptoms [9,27]. Similar stimulation parameters were used for posterior tibial nerve stimulation [4,28]. Five patients in this study could only tolerate stimulatory amplitudes at or below 11 mA, which was less than two times the clitoro-anal reflex threshold. Even so, a reduction in the number of bowel movements, urgency episodes, and urgency incontinence episodes was observed in two of those three patients. This indicates that amplitudes lower than two times the reflex threshold are sufficient to treat FI. With this lower amplitude, stimulation is more tolerable, and patient compliance is likely to increase. The protocol, which involved stimulation twice daily for three weeks was chosen arbitrarily. A high frequency of stimulation sessions was chosen to test the feasibility of the concept. The standard three-week bowel habit diary used in our unit was adapted, to avoid any uncertainty due to fluctuations in bowel habits. There appeared to be an immediate response in a number of patients (Table 1), and therefore, it would be interesting to investigate whether a three-week protocol is actually necessary to improve symptoms.

Two studies have explored DGN stimulation as a treatment for FI. Binnie et al. initially demonstrated improved function of the external anal sphincter and continence in 7 of 8 patients after five minutes of stimulation three times a day for eight weeks [10]. The mean stimulation amplitude was 135 V (standard deviation: 15 V), which was two to three times higher than the sensory threshold. The pulse width was 0.1 ms and the pulse rate was 1 Hz. Frizelle et al. reported improved Wexner incontinence scores accompanied by significant improvements in anal sphincter tone and maximal tolerated rectal volume following stimulation three times a day for eight weeks with self-titrated voltage from 30 to 90 V using a pulse width of 0.1 ms and a pulse rate of 2 Hz [11]. The improved sphincter function seen in both studies was most likely due to repeated low frequency DGN stimulation which intermittently trains the muscles for a period of eight weeks. With higher stimulation frequencies this is not the case, as the reflex response disappear after a short time with stimulation at for example 20 Hz. Therefore, the effects of stimulation in this study must have been due to a different mode of action, and treatment for eight weeks might not be necessary.

Stimulation of the posterior tibial nerve to treat FI has also been investigated and is currently in used. Stimulation is performed as an ambulatory procedure with scheduled sessions at different intervals [4,28]. Stimulation is performed using transcutaneous plaster electrodes or percutaneous stimulation with a needle electrode. A pulse width of 0.2 ms and pulse rates from 10 to 20 Hz with variable amplitudes are used. In these studies, symptoms decreased with stimulation, whereas the effects on sphincter function were variable. Shafik et al. applied stimulation every other day for four weeks and reported improvement of the Wexner score and the rectal pressure-volume relation [28]. De la Portilla et al. reported improvement of 10 out of 16 patients, and Boyle et al. reported improvement in 21 out of 31 patients; both studies used weekly stimulation for 12 weeks, which is comparable to our results [4,29]. Furthermore, after the initial treatment, maintenance therapy with less frequent stimulation sessions appears to be sufficient to sustain the reduction in symptoms. It is possible, that fewer stimulation sessions are sufficient with DGN stimulation as well.

The mechanism of action for DGN stimulation is unknown. Vitton et al. showed that peripheral stimulation could invoke a somatosympathetic reflex [30]. In acute experiments with cats, both stimulation of somatic (sciatic and radial) nerves and sacral roots (S1-S3) with a pulse width of 210 μs at a pulse rate of 10 Hz have exibited an inhibitory effect on colonic electromyographic activity and enhanced activity of the internal anal sphincter. It is unknown whether similar effects occur with DGN stimulation.

A placebo effect cannot be ruled out in the present study mainly because the patient's key problem was addressed, with repeated investigations and attention from health staff. This effect might be supported by the fact that patients who were stimulated with very low amplitude also experienced some improvement. A randomised double-blinded setup, preferably with sham stimulation is necessary to determine whether this is the case. A larger study should be initialised to confirm this result. A definition of successful treatment similar to 50% improvement during the test period used for sacral nerve stimulation should be established. Whether the initial reduction in symptoms is maintained over a longer follow-up period also needs to be clarified. Furthermore, optimal stimulation parameters and intervals between stimulations should be defined. In that case, peripheral stimulation could represent a cheaper and less invasive alternative to surgical treatments including sacral nerve stimulation.

It must be recognised that clitoral stimulation might be unacceptable to some patients. Even so, in our experience this treatment can be introduced without any difficulty by educating patients about the procedure and training by a dedicated nurse.

2.5 CONCLUSION

Electrical DGN stimulation twice daily for three weeks reduces incontinence episodes in patients suffering from idiopathic FI. The reduction is maintained for three weeks after stimulation. No improvement in disease-specific quality of life was found. No changes in sphincter function or rectal volume tolerability were observed. DGN stimulation may represent a new treatment for FI.

REFERENCES

[1]   Pretlove SJ, Radley S, Toozs-Hobson PM, Thompson PJ, Coomarasamy A, Khan KS. Prevalence of anal incontinence according to age and gender: a systematic review and meta-regression analysis. Int Urogynecol J Pelvic Floor Dysfunct 2006;17:407-1.

[2]   Miner PB,Jr. Economic and personal impact of fecal and urinary incontinence. Gastroenterology 2004;126:S8-13.

[3]   National Institute for Health and Clinical Excellence. Faecal incontinence: the management of faecal incontinence in adults (CG49). 2007.

[4]   de la Portilla F, Rada R, Vega J, Gonzalez CA, Cisneros N, Maldonado VH. Evaluation of the use of posterior tibial nerve stimulation for the treatment of fecal incontinence: preliminary results of a prospective study. Dis Colon Rectum 2009;52:1427-33.

[5]   Melenhorst J, Koch SM, Uludag Ö, Van Gemert WG, Baeten CG. Sacral neuromodulation in patients with faecal incontinence: Results of the first 100 permanent implantations. Colorectal Disease 2007;9:725-30.

[6]   Michelsen HB, Thompson-Fawcett M, Lundby L, Krogh K, Laurberg S, Buntzen S. Six years of experience with sacral nerve stimulation for fecal incontinence. Dis Colon Rectum 2010;53:414-21.

[7]   Vitton V, Damon H, Roman S, Mion F. Posterior tibial nerve stimulation and fecal incontinence. Hepato-Gastro 2009;16:191-4.

[8]   Goldman HB, Amundsen CL, Mangel J, Grill J, Bennett M, Gustafson KJ et al. Dorsal genital nerve stimulation for the treatment of overactive bladder symptoms. Neurourol Urodyn 2008;27:499-503.

[9] Hansen J, Media S, Nohr M, Biering-Sorensen F, Sinkjaer T, Rijkhoff NJ. Treatment of neurogenic detrusor overactivity in spinal cord injured patients by conditional electrical stimulation. J Urol 2005;173:2035-9.

[10] Binnie NR, Kawimbe BM, Papachryosostomou M, Smith AN. Use of the pudendo-anal reflex in the treatment of neurogenic faecal incontinence. Gut 1990;31:1051-5.

[11] Frizelle FA, Gearry RB, Johnston M, Barclay ML, Dobbs BR, Wise C et al. Penile and clitoral stimulation for faecal incontinence: External application of a bipolar electrode for patients with faecal incontinence. Colorectal Disease 2004;6:54-7.

[12] Jorge JM, Wexner SD. Etiology and management of fecal incontinence. Dis Colon Rectum 1993 36:77-9.

[13] Vaizey CJ, Carapet E, Cahill JA, Kamm MA. Prospective comparison of faecal incontinence grading systems. Gut 1999;44:77-80.

[14] Rockwood TH, Church JM, Fleshman JW, Kane RL, Mavrantonis C, Thorson AG et al. Fecal Incontinence Quality of Life Scale: quality of life instrument for patients with fecal incontinence. Dis Colon Rectum 2000;43:9,16; discussion 16-7.

[15] Maeda Y, Pares D, Norton C, Vaizey CJ, Kamm MA. Does the St. Mark's incontinence score reflect patients' perceptions? A review of 390 patients. Dis Colon Rectum 2008;51:436-42.

[16] Deutekom M, Terra MP, Dobben AC, Dijkgraaf MG, Felt-Bersma RJ, Stoker J et al. Selecting an outcome measure for evaluating treatment in fecal incontinence. Dis Colon Rectum 2005;48:2294-301.

[17] Ryhammer AM, Laurberg S, Hermann AP. Test-retest repeatability of anorectal physiology tests in healthy volunteers. Dis Colon Rectum 1997;40:287-92.

[18] McHugh SM, Diamant NE. Effect of age, gender, and parity on anal canal pressures. Contribution of impaired anal sphincter function to fecal incontinence. Dig Dis Sci 1987;32:726-3.

[19] Fernandez-Fraga X, Azpiroz F, Malagelada JR. Significance of pelvic floor muscles in anal incontinence. Gastroenterology 2002;123:1441-50.

[20] Matzel KE, Stadelmaier U, Hohenfellner M, Gall FP. Electrical stimulation of sacral spinal nerves for treatment of faecal incontinence. Lancet 1995;346:1124-7.

[21] Vaizey CJ, Kamm MA, Turner IC, Nicholls RJ, Woloszko J. Effects of short term sacral nerve stimulation on anal and rectal function in patients with anal incontinence. Gut 1999;44:407-12.

[22]   Vaizey CJ, Kamm MA, Roy AJ, Nicholls RJ. Double-blind crossover study of sacral nerve stimulation for fecal incontinence. Dis Colon Rectum 2000;43:298-302.

[23]   Ganio E, Ratto C, Masin A, Luc AR, Doglietto GB, Dodi G et al. Neuromodulation for fecal incontinence: outcome in 16 patients with definitive implant. The initial Italian Sacral Neurostimulation Group (GINS) experience. Dis Colon Rectum 2001;44:965-70.

[24]   Tjandra JJ, Chan MK, Yeh CH, Murray-Green C. Sacral nerve stimulation is more effective than optimal medical therapy for severe fecal incontinence: a randomized, controlled study. Dis Colon Rectum 2008;51:494-502.

[25]   Leroi AM, Parc Y, Lehur PA, Mion F, Barth X, Rullier E et al. Efficacy of sacral nerve stimulation for fecal incontinence: Results of a multicenter double-blind crossover study. Ann Surg 2005;242:662-9.

[26]   Michelsen HB, Buntzen S, Krogh K, Laurberg S. Rectal volume tolerability and anal pressures in patients with fecal incontinence treated with sacral nerve stimulation. Dis Colon Rectum 2006;49:1039-44.

[27]   Previnaire JG, Soler JM, Perrigot M, Boileau G, Delahaye H, Schumacker P et al. Short-term effect of pudendal nerve electrical stimulation on detrusor hyperreflexia in spinal cord injury patients: importance of current strength. Paraplegia 1996;34:95-9.

[28]   Shafik A, Ahmed I, El-Sibai O, Mostafa RM. Percutaneous peripheral neuromodulation in the treatment of fecal incontinence. Eur Surg Res 2003;35:103-7.

[29]   Boyle DJ, Prosser K, Allison ME, Williams NS, Chan CL. Percutaneous tibial nerve stimulation for the treatment of urge fecal incontinence. Dis Colon Rectum 2010;53:432-7.

[30]   Vitton V, Abysique A, Gaigé S, Leroi AM, Bouvier M. Colonosphincteric electromyographic responses to sacral root stimulation: Evidence for a somatosympathetic reflex. Neurogastroenterology and Motility 2008;20:407-16.

Chapter   3.

The acute effect of dorsal genital nerve stimulation on rectal wall properties in patients with idiopathic faecal incontinence

J. Worsøe[#], L. Fynne[+], S. Laurberg[*] K. Krogh[+], N.J.M. Rijkhoff[#]

[#]*Center for Sensory-Motor Interaction (SMI), Department of Health Science and Technology, Aalborg University, Denmark*
[*]*Department of Surgery P, Aarhus University Hospital, Aarhus, Denmark*
[+]*Neurogastroenterology Unit, Department of Hepatology and Gastroenterology V, Aarhus University Hospital, Aarhus, Denmark*

3.1 INTRODUCTION

Faecal incontinence (FI) is a common symptom with a prevalence of 4.3% in the general population [1]. There are various aetiologies of FI, including sphincter damage, anorectal disease, sequelae from anorectal surgery, neural lesions, and systemic disease. Idiopathic FI accounts for greater than 50% of patients with FI, and there is no obvious aetiology [2]. Treatment for FI is usually offered in a stepwise manner, starting with advice about diet and lifestyle, followed by constipating agents and biofeedback training. If the sphincter is damaged, surgical repair can be performed, but long-term results are often poor [3,4]. More extensive surgical procedures include artificial sphincter or stoma formation.

Electrical nerve stimulation is also used to treat FI. Stimulation of the sacral roots (S2-S4) using Interstim Therapy has been shown to reduce FI symptoms and improve quality of life in approximately 80% of the patients given this treatment in previous studies [5,6]. The surgical procedure for electrical nerve stimulation is minimally invasive, and the response to stimulation can be tested prior to permanent electrode implantation [7]. Stimulation of the posterior tibial nerve (Urgent PC) is another approach that does not require surgery. Stimulation sessions of the posterior tibial nerve once or twice a week reduce FI symptoms [8,9].

The dorsal genital nerve (DGN) may represent an alternative target for peripheral nerve stimulation treatment of FI. DGN stimulation inhibits bladder contractions and decreases filling pressure [10,11]. Because the bladder and rectum have similar innervation, an equivalent inhibitory effect on the rectum was expected. A study in spinal cord injury (SCI) patients demonstrated increased rectal compliance during DGN stimulation [12]. Two studies reported a reduction in FI symptoms and an improvement of sphincter function with DGN stimulation in patients with FI [13,14].

FI is influenced by several factors, including sphincter function, anorectal sensibility, colorectal motility, rectal capacity, and load and consistency of bowel content [15]. Rectal capacity depends both on the size of the rectum and on rectal wall properties. A compliant rectum has a high tolerable volume, which lowers the rectal pressure, so that the pressure of the sphincter apparatus is not overcome. Rectal compliance is typically measured with pressure-volume based techniques using balloons for distension. However, another method, rectal impedance planimetry (IP), determines rectal cross-sectional area (CSA), and a pressure-CSA relationship can be derived with simultaneous measurement of rectal pressure. Thus, with IP, the errors associated with volume-based methods are avoided [16-19]. The aim of the present study was to determine whether DGN stimulation has an acute effect on the rectal CSA in patients with idiopathic FI similar to what has been demonstrated in SCI patients [12].

3.2 METHODS

3.2.1   Subjects

Ten female patients with idiopathic FI were included in the study (median age: 60 years; range: 34 to 68 years). Initial assessment included full medical history, anorectal physiology testing, anal ultrasonography, colonic transit time, and anorectoscopy. Patients with significant sphincter lesions, previous pelvic or anorectal surgery, organic anorectal disease, and systemic diseases affecting bowel function or neurological disorders were excluded. During the investigation, the patients lay in the left lateral position and were asked to remain as still as possible and to refrain from talking. The study was performed in accordance with the Helsinki Declaration II and was approved by the Ethical Committee of Region Midtjylland (M-20090145). All participants provided informed written consent.

3.2.2 Stimulation

Stimulation was performed with a constant-current stimulator (Digitimer model DS7A, Digitmer Ltd., Welwyn Garden City, England), and the frequency was controlled with a function generator (Agilent 33120A, Hewlett-Packard, USA) to deliver square pulses. A pulse width of 200 µs and a frequency of 20 Hz were used. The amplitude was set at the highest level that the patient could tolerate (preferably two times the pudendo-anal reflex threshold). One electrode (dimensions: 10 x 20 mm, Neuroline 700, Ambu, Ballerup, Denmark) was placed on the clitoris (cathode), and a second electrode (diameter: 32 mm, PALS Platinum, Axelgaard, Lystrup, Denmark) was placed 2–3 cm lateral to the right labia major (fig. 2.1). Precautions were taken to ensure good contact between the skin and the electrodes. The clitoro-anal reflex was identified visually during slow stepwise increases of the stimulation amplitude before starting the rectal distensions.

3.2.3 Impedance planimetry

IP was used to determine the rectal CSA (fig. 3.1). The CSA is proportionally related to the potential difference between two detection electrodes in a uniform electrical field. Given a current I, the potential difference between two detection electrodes in a fluid with the conductivity σ, is $\Delta V = I\, d\, \sigma\text{-}1 CSA\text{-}1$, where CSA is the luminal cross-sectional area and d is the distance between two detection electrodes. A custom-made reusable probe was used. The probe had two excitation electrodes 60 mm apart, providing a sinusoidal current of 0.1 mA at 10 kHz, and one pair of detection electrodes (3 mm apart), situated 40 mm above the anal canal, for measurement of rectal CSA. The electrodes were within a non-compliant flaccid bag (dimensions: 90x90 mm), which was filled with 0.9% saline at 37°C. The pressure within the bag was controlled by elevation of an open water container. Before starting the measurements, a multipoint calibration was performed using circular tubes with CSAs from 233 to 4322 mm2, which was the range of the expected CSAs. Intraluminal rectal pressure and anal canal pressure were measured using perfused catheters within the bag and within a balloon placed in the anal canal, both of which were connected to pressure transducers. The pressure transducers were calibrated using 0 and 100 cm H2O as the minimum and maximum. All signals were sampled at 10 Hz, and the data were visualised and stored using custom-made software (Openlab, Noerresundby, Denmark). Prior to placement of the probe, resting rectal pressure was measured with a water-perfused catheter placed in the rectum. The equipment and method used for IP have been described previously [20,21].

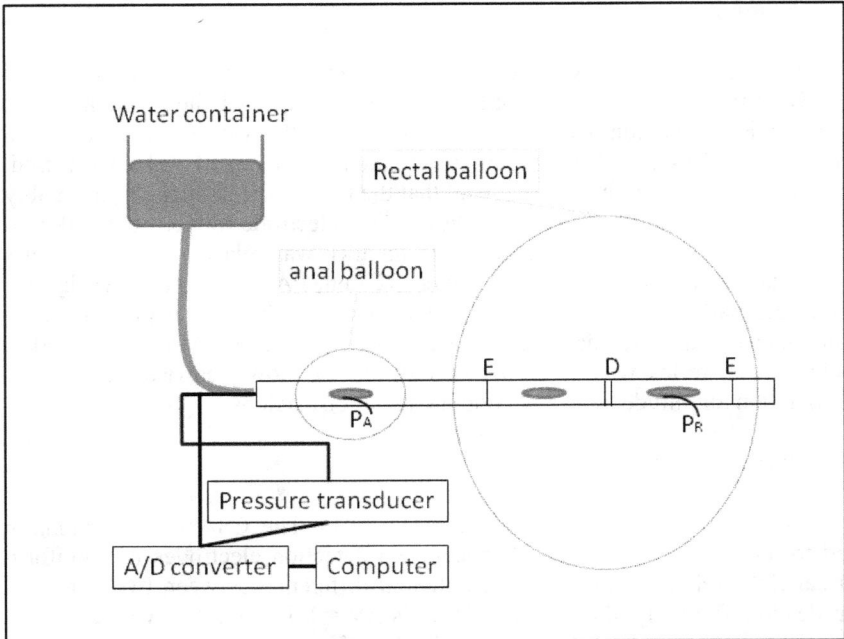

Figure 3.1: A schematic illustration of the system for impedance planimetry. E: excitation electrode, D: detection electrodes, PA: anal pressure, PR: rectal pressure. The CSA was measured at the level of the detection electrodes, and all signals were A/D converted for storage and computer visualisation.

3.2.4   Distension protocol

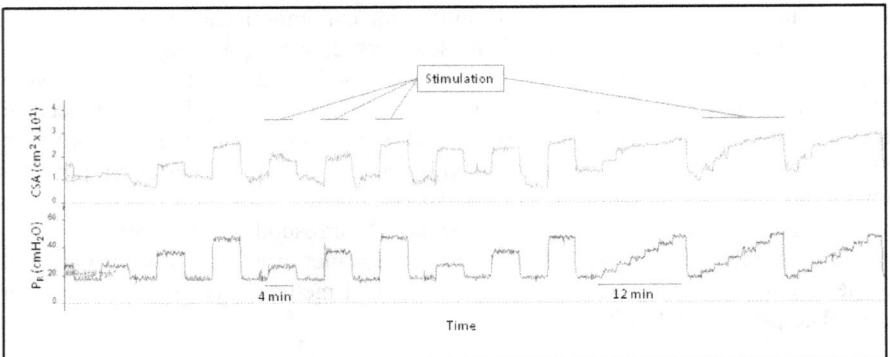

Figure 3.2: An example of a distension series from an experimental recording. Three phasic distension series (10, 20, and 30 cm H2O) were performed, followed by three stepwise distensions (5, 10, 15, 20, 25, and 30 cm H2O). Distension at each level of the stepwise distension lasted 2 min. CSA: cross-sectional area, PR: rectal pressure.

A pressure-controlled combined phasic and stepwise distension protocol was performed (fig. 3.2). The total length of the experiment was approximately 150 min, including the preparation and placement of the probe. The experiment started with a rectal distension of 10 cm H_2O above resting rectal pressure for 12 min, which served as a conditioning distension. The conditioning distension was followed by four minutes of distension at the resting rectal pressure to measure the resting rectal CSA. Thereafter, phasic distensions at 3 pressure levels (10, 20, and 30 cm H_2O) were conducted. The distensions lasted for four minutes and were separated by four minutes with the pressure lowered to the resting rectal pressure. This distension sequence was repeated three times (control 1, stimulation, control 2), and during the second sequence, DGN stimulation was applied during the distensions. After 4 min of distension at the resting rectal pressure, three series of pressure-controlled stepwise distensions (5, 10, 15, 20, 25, and 30 cm H_2O) were performed with each step lasting two minutes and with DGN stimulation applied during the second stepwise distension.

3.2.5   Analysis

The rectal CSA and the rectal pressure were determined when a steady state was achieved at each pressure level. Steady state rectal CSA, calculated as the mean CSA during the last minute of the distension, was assumed when the difference in the mean CSA between the first 10 s period and the last 10 s period was less than 10% (fig. 3.3). The rectal pressure-CSA relationship, "CSA"/"PR" , was calculated for each pressure level. The circumferential wall tension was calculated using Laplace's law, $T = \Delta p\, r$, where T is the circumferential wall tension, r is the radius of the rectum and Δp is the transmural pressure calculated as the difference between the resting rectal pressure P0 and the pressure during distension PR.

3.2.6   Statistics

Numerical data are expressed as medians with ranges. Comparisons between the baseline values and the stimulation and post stimulation values were conducted using the Wilcoxon's test for non-parametric comparisons of paired measurements.

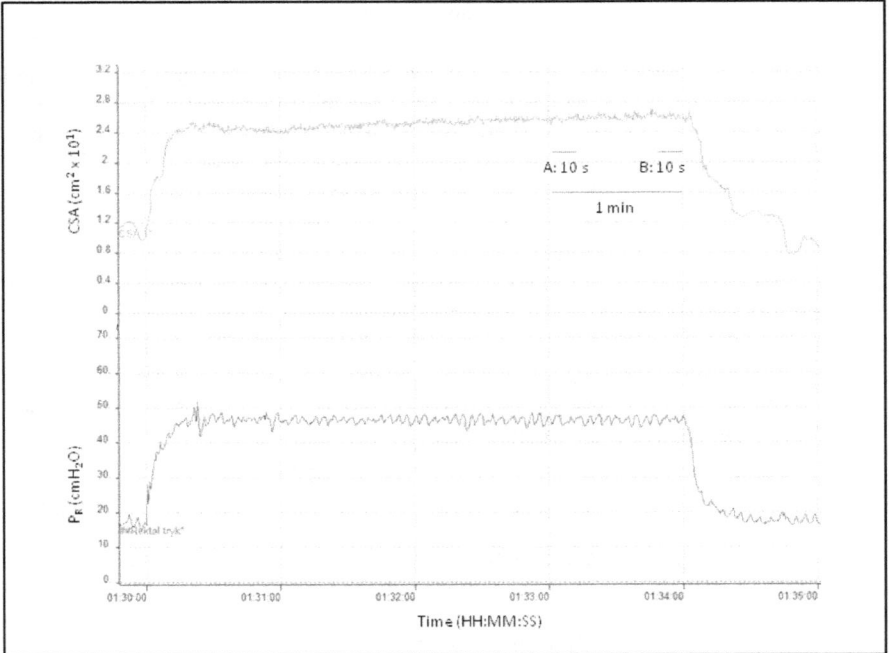

Figure 3.3: Determination of the rectal CSA (example). The mean CSA for a 1 min period was valid if the mean CSA for the first 10 s (A) and the last 10 s (B) differed by less than 10%. CSA: rectal cross-sectional area, PR: rectal pressure.

3.3 RESULTS

All patients had normal colonic transit times (<5 of the markers were present after six days). Transanal ultrasonography demonstrated intact sphincters in eight patients and minor defects in two patients. The anal squeeze pressure, anal resting pressure, rectal volume tolerability, left and right pudendal nerve terminal motor latency (PNTML) and anal sensation were normal in all except one patient, who could not produce any voluntary squeeze despite normal PNTML and normal sphincter configuration. All patients tolerated the investigation well, and there were no adverse events. The median clitoro-anal reflex threshold, median rectal pressure at rest, median rectal CSA at resting rectal pressure, and median stimulation amplitude were 9 mA (range: 8-25 mA), 9.5 cmH2O (range: 8-16 cmH2O), 15 cm2 (range 9 to 30), and 21 mA (range 8.5-27). During the experiment, stimulations above the clitoro-anal reflex threshold were observed as brief (1-3 s) increases in anal pressure. Filling of the rectal bag resulted in an increase in rectal CSA. Most of that increase occurred within the first 30 s, and thereafter, the rectal CSA became constant, with changes of less than 10% during the fourth minute in all of the patients.

3.3.1   Phasic distension

Data are shown in figure 3.4 and in table 3.1. Data from one patient were discarded due to problems with the recording system. The median rectal CSA increased with increasing rectal distension pressure. With distensions at 10, 20, and 30 cmH2O above resting rectal pressure, there were no significant differences between rectal CSA obtained before and during DGN stimulation (control 1 vs. stimulation), or between rectal CSA before and after DGN stimulation (control 1 vs. control 2). During treatment with 10 cmH2O of distension pressure, the rectal CSA was higher during the DGN-stimulated distension compared to control 1 and control 2 in three patients only. Two of those patients tolerated high stimulation amplitudes (21 and 25 mA), whereas the third patient tolerated only a low stimulation amplitude (8.5 mA). At 20 cmH2O of distension pressure, only one patient had a higher rectal CSA during stimulation, and at 30 cmH2O of distension pressure, no difference in the rectal CSA was observed during stimulation in any of the patients. Therefore, there were no significant differences between the rectal pressure-CSA relationship before, during, or after stimulation at any of the distension pressures. The rectal wall tension increased with rectal distension pressure. Similarly, there were no significant differences between the rectal wall tension before, during, or after stimulation at any of the distension pressures.

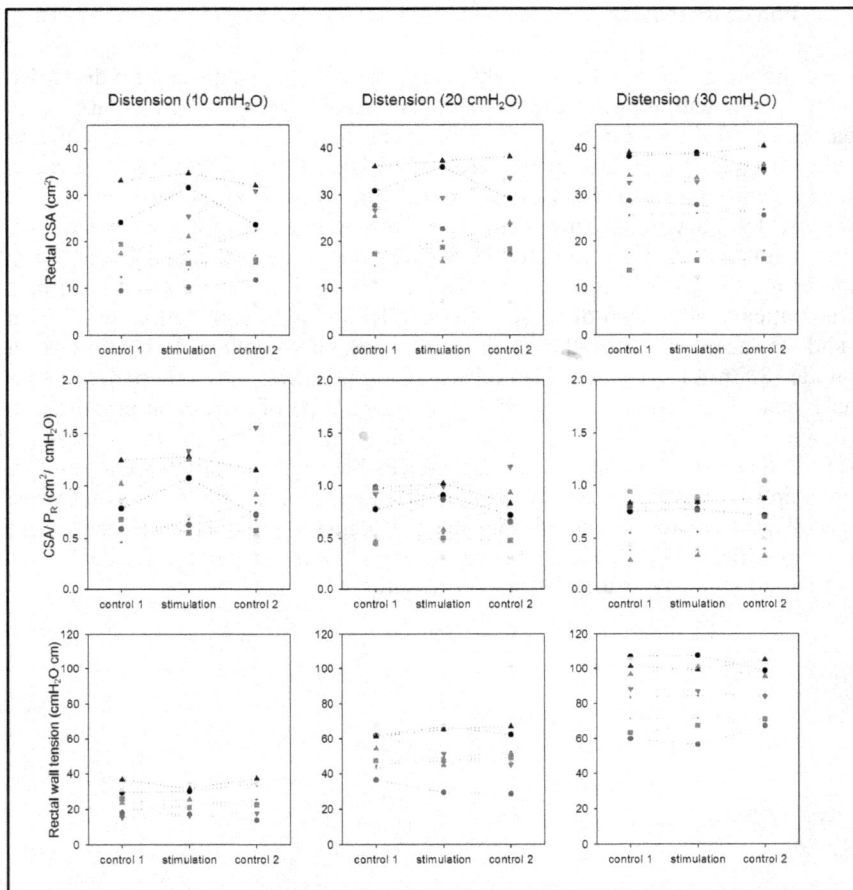

Figure 3.4: Phasic distension. Data from nine patients are shown. Unstimulated (control 1 and control 2) and stimulated distensions are compared at different distension pressures (10, 20, and 30 cm H2O). Rectal CSA, rectal CSA-pressure relation (CSA/PR), and rectal wall tension are shown.

		Rectal distension pressure (cmH$_2$O)		
		10	20	30
Rectal CSA (cm^2)	Control 1	17 (9-33)	24 (11-36)	31 (14-39)
	Stimulation	17 (9-35)	19 (7-37)	27 (12-39)
	Control 2	17 (7-32)	21 (7-42)	26 (15-40)
CSA/P$_R$ (cm^2 / cmH$_2$O)	Control 1	0.74 (0.46-1.24)	0.73 (0.42-0.99)	0.77 (0.29-0.94)
	Stimulation	0.70 (0.50-1.33)	0.57 (0.30-1.02)	0.68 (0.33-0.89)
	Control 2	0.73 (0.49-1.55)	0.65 (0.23-1.18)	0.64 (0.33-1.04)
Wall tension (cmH$_2$O cm)	Control 1	24 (16-37)	52 (36-67)	92 (60-107)
	Stimulation	22 (17-34)	48 (29-68)	86 (57-108)
	Control 2	23 (14-37)	50 (28-101)	84 (64-105)

Table 3.1: Rectal CSA, rectal CSA/PR, and rectal wall tension during the phasic distension protocol. Three consecutive distension series (10, 20, and 30 cm H2O), unstimulated (control 1), stimulated, and unstimulated (control 2), are shown. The data are expressed in the following form: median (range).

3.3.2   Stepwise distension

Data from the stepwise distensions are presented in figure 3.5. In all of the distension series, the median rectal CSA increased during the stepwise distension from 5 to 30 cmH2O. There were no significant differences in the rectal CSA between control 1, stimulation, and control 2. The ratio of the rectal pressure-CSA relationship decreased with increasing distension pressure. There were no significant differences in the pressure-CSA relationship between control 1, stimulation, or control 2. Rectal wall tension increased linearly with increasing distension pressure (fig. 3.5). There were no significant changes in the rectal wall tension when control 1 was compared to the stimulated distension or control 2.

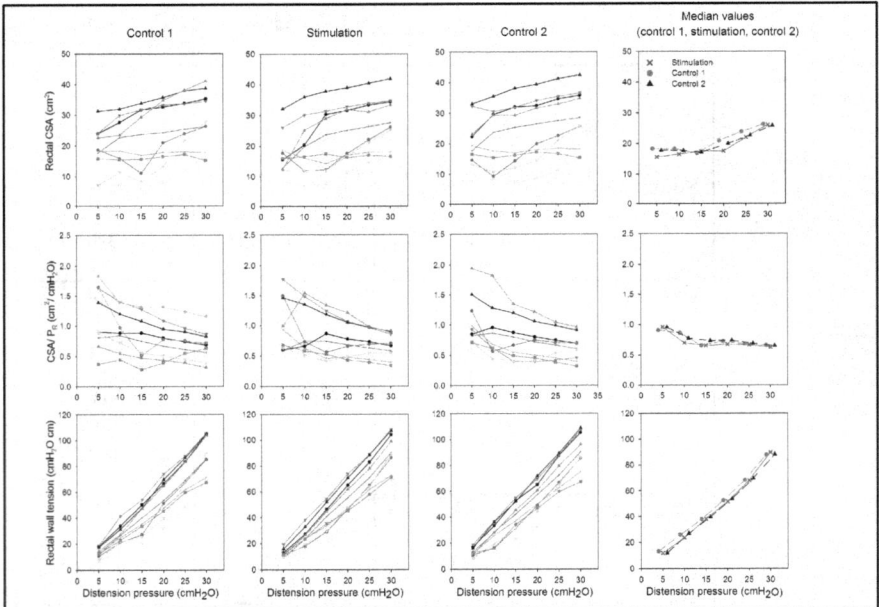

Figure 3.5: Stepwise distension (5–30 cm H2O, in steps of 5 cm H2O). Data from ten patients and medians are shown. Rectal CSA, the rectal pressure-CSA relation (CSA/ PR), and rectal wall tension are shown during rectal distension before (control 1), after (control 2) and during DGN stimulation.

3.4 DISCUSSION

The aim of this study was to investigate the effect of DGN stimulation on rectal CSA during distension with an inflatable bag. No acute effects on the rectal CSA, the rectal wall tension or the rectal pressure-CSA relationship were demonstrated. Experiments that involved inhibition with glucagon demonstrated that a basal tone exists in the rectum [22]. It was expected that DGN stimulation would inhibit the smooth muscles of the rectum and increase the rectal CSA during stimulated distension compared to the before and after control distensions.

The stimulation parameters were based on previous experiences with DGN stimulation used to achieve inhibition of the bladder, where a pulse width of 200 μs and a pulse rate of 20 Hz with an amplitude of at least twice the clitoro-anal reflex threshold were used [11,23]. With a similar stimulation amplitude (a frequency of 20 Hz and a pulse width of 200 μs), DGN stimulation has also been used to induce acute inhibition of the rectum in SCI patients [12]. In that SCI study, different stimulation frequencies were also tested (0.2, 2, and 20 Hz), and rectal compliance was highest during the 20 Hz stimulation. In the present study, the patients were stimulated at the maximum tolerable amplitude, which was a median of 30% (range: 6% to 50%) above the clitoro-anal reflex threshold. None of the patients could tolerate a stimulation amplitude of twice the reflex threshold, and only three patients tolerated stimulation amplitudes above 25 mA. Five of the patients could only tolerate stimulation amplitudes 1–3 mA above the reflex threshold. Therefore, it is possible that a lack of an acute response on the rectal CSA was observed because the stimulation amplitude was too low. Nevertheless, DGN stimulation at a low amplitude may still reduce FI symptoms. Nine of the 10 patients from the present study were later treated with the maximum tolerable DGN stimulation twice a day for three weeks. FI symptoms were reduced in seven of the nine patients (in press). This finding is consistent with data on bladder stimulation used to treat urinary incontinence. Goldman et al. used DGN stimulation to inhibit bladder overactivity, assessing both cystometry during acute stimulation and urinary incontinence symptoms during 1 week of home stimulation [10]. The mean stimulation amplitude was 9.6 ± 4.9 mA (range: 4–24 mA). No acute response was observed with cystometry during stimulation, while there was a significant reduction of urinary incontinence symptoms during home stimulation.

The range of measured rectal CSAs was from 7 to 41 cm2, with large variation between the patients. During phasic distensions, the variation in rectal CSA within each patient was larger at the 10 cm H2O distension pressure than at the 30 cm H2O distension pressure. However, most patients with a large CSA during control 1 also had a large CSA during control 2 and vice versa. At 10 cm H2O, the median difference in CSA between control 1 and control 2 was 2% (range: -80% to 40%). At 30 cm H2O distension pressure, the median difference in CSA was 5% (range: -13% to 11%). The CSA measurements were comparable with findings in studies using IP in healthy volunteers. Dall et al. measured median rectal CSAs from approximately 7 cm2 at 5 cm H2O distension pressure to 35 cm2 at 30 cm H2O distension pressure, and Krogh et al. found mean rectal CSAs from approximately 10 cm2 at 10 cm H2O distension pressure to 35 cm2 at 30 cm H2O distension pressure [16,24]. At present, no data on rectal CSA during distension exists in patients with idiopathic FI, but a large overlap with healthy subjects is expected. Rectal volume tolerability assessed with balloon distension appears to be lower in some FI patients compared to healthy subjects, as they had lower rectal volume thresholds for first sensation [25,26].

The acute effects of DGN stimulation on rectal wall properties have been investigated in SCI patients but have never been reported in non-neurogenic patients. Chung et al. investigated the acute effect of DGN stimulation in SCI patients with suprasacral lesions [12]. In their study, a rectal barostat caused distension, with and without electrical stimulation at twice the pudendo-anal reflex threshold, and the rectal compliance was studied. The distensions were performed in 50 ml steps, up to 300 ml. The maximum rectal compliance was approximately 7 ml/cm H2O without stimulation, and 12 ml/cm H2O during stimulation. The rectal compliances with and without stimulation were both within the range of normal rectal compliance reported in the literature, which ranges from 4.5 to 17 ml/cm H2O [27,28]. However, several studies have reported large variations between the individual patients and also between the mean values reported by different institutions when measuring the relationship between pressure and volume in the rectum, which makes comparison with other studies difficult [19,29,30]. Even so, rectal compliance or the rectal pressure-CSA relation may be important in the pathophysiology of FI. It has recently been shown that the postprandial rectal tone is enhanced in patients with idiopathic FI compared to healthy controls and that successful treatment with InterStim Therapy reduced postprandial rectal tone in these patients [31,32]. Rectal wall tension may be related to activation of sensory receptors and conscious feeling of distension [33]. Because of the simultaneous DGN stimulation, sensory thresholds were not assessed in this study. Any effect of stimulation that changed the rectal CSA would also have affected the rectal wall tension.

It has been demonstrated that the configuration of the distension profile has little effect on distensibility [34]. Both phasic and stepwise distensions were performed in this experiment to study the rectum during different loading patterns. Phasic distension simulates the sudden arrival of faecal content with an increase in rectal pressure, while the stepwise distension exposes the rectum to a longer lasting, progressively increased pressure load. Distensions lasting four minutes were used in this investigation. In this first-order system, the change in CSA per unit time will continue to decrease, and the decision to calculate rectal CSA based on the average of the fourth minute would affect the reported CSA values. A longer distension period would produce a more stable and higher CSA, but this advantage had to be balanced against other limitations. Such prolonged distensions might affect mucosal blood flow at high distension pressures, and the total length of the experiment had to be acceptable to the patients. Nevertheless, the length of the distension periods is unlikely to have influenced the conclusions, because the changes in CSA were small (less than 10%) and comparisons were made between CSAs calculated using the same time period during the distension.

The signal conditioning system did not include a low-pass filter (anti-aliasing). However, this is not expected to influence the results because no events of interest occurred at high frequencies and the signal noise had a low amplitude compared with the changes in the rectal CSA. While the rectal CSA measurements during distensions were reproducible, the CSAs at basic rectal pressure were not stable. At low distension pressures, the bag tended to fold and become irregularly shaped, which affected the measurements. Therefore, reliable CSA measurements can only be obtained using IP during distension above the resting rectal pressure. An eccentric position of the probe could also affect measurements. Correct positioning of the probe has been confirmed with ultrasound previously [35].

3.5 CONCLUSSION

In conclusion, no acute effects on rectal CSA due to DGN stimulation during pressure-controlled distensions were found. Based on this investigation, it can be concluded that it is not always possible to use stimulation amplitudes near two times the clitoro-anal reflex threshold in non-neurogenic patients.

REFERENCES

[1]   Pretlove SJ, Radley S, Toozs-Hobson PM, Thompson PJ, Coomarasamy A, Khan KS. Prevalence of anal incontinence according to age and gender: a systematic review and meta-regression analysis. Int Urogynecol J Pelvic Floor Dysfunct 2005;17:407-1.

[2]   Ilnyckyj A. Prevalence of idiopathic fecal incontinence in a community-based sample. Can J Gastroenterol 2010;24:251-4.

[3]   Malouf AJ, Norton CS, Engel AF, Nicholls RJ, Kamm MA. Long-term results of overlapping anterior anal-sphincter repair for obstetric trauma. Lancet 2000;355:260-5.

[4]   Vaizey CJ, Norton C, Thornton MJ, Nicholls RJ, Kamm MA. Long-term results of repeat anterior anal sphincter repair. Dis Colon Rectum 2004;47:858-63.

[5]   Leroi A-, Parc Y, Lehur P-, Mion F, Barth X, Rullier E et al. Efficacy of sacral nerve stimulation for fecal incontinence: Results of a multicenter double-blind crossover study. Ann Surg 2005;242:662-9.

[6]   Melenhorst J, Koch SM, Uludag Ö, Van Gemert WG, Baeten CG. Sacral neuromodulation in patients with faecal incontinence: Results of the first 100 permanent implantations. Colorectal Disease 2007;9:725-30.

[7]   Matzel KE, Kamm MA, Stosser M, Baeten CGMI, Christiansen J, Madoff R et al. Sacral spinal nerve stimulation for faecal incontinence: Multicentre study. Lancet 2004;363:1270-5.

[8]   de la Portilla F, Rada R, Vega J, Gonzalez CA, Cisneros N, Maldonado VH. Evaluation of the use of posterior tibial nerve stimulation for the treatment of fecal incontinence: preliminary results of a prospective study. Dis Colon

Rectum 2009;52:1427-33.

[9]   Vitton V, Damon H, Roman S, Mion F. Posterior tibial nerve stimulation and fecal incontinence. Hepato-Gastro 2009;16:191-4.

[10]   Goldman HB, Amundsen CL, Mangel J, Grill J, Bennett M, Gustafson KJ et al. Dorsal genital nerve stimulation for the treatment of overactive bladder symptoms. Neurourol Urodyn 2008;27:499-503.

[11]   Hansen J, Media S, Nohr M, Biering-Sorensen F, Sinkjaer T, Rijkhoff NJ. Treatment of neurogenic detrusor overactivity in spinal cord injured patients by conditional electrical stimulation. J Urol 2005;173:2035-9.

[12]   Craggs MD, Balasubramaniam AV, Chung EA, Emmanuel AV. Aberrant reflexes and function of the pelvic organs following spinal cord injury in man. Auton Neurosci 2006;126-127:355-70.

[13]   Binnie NR, Kawimbe BM, Papachryosostomou M, Smith AN. Use of the pudendo-anal reflex in the treatment of neurogenic faecal incontinence. Gut 1990;31:1051-5.

[14]   Frizelle FA, Gearry RB, Johnston M, Barclay ML, Dobbs BR, Wise C et al. Penile and clitoral stimulation for faecal incontinence: External application of a bipolar electrode for patients with faecal incontinence. Colorectal Disease 2004;6:54-7.

[15]   Baeten C, Kuijpers HC. Incontinence in The ASCRS Textbook of Colon and Rectal Surgery (Editors: Wolff, B.G.; Fleshman, J.W.; Beck, D.E.; Pemberton, J.H.; Wexner, S.D.). 2007:pp 653-664.

[16]   Dall FH, Jorgensen CS, Houe D, Gregersen H, Djurhuus JC. Biomechanical wall properties of the human rectum. A study with impedance planimetry. Gut 1993;34:1581-6.

[17]   Krogh K, Ryhammer AM, Lundby L, Gregersen H, Laurberg TS. Comparison of methods used for measurement of rectal compliance. Dis Colon Rectum 2001;44:199-206.

[18]   Lundby L, Krogh K, Jensen VJ, Gandrup P, Qvist N, Overgaard J et al. Long-term anorectal dysfunction after postoperative radiotherapy for rectal cancer. Dis Colon Rectum 2005;48:1343,9; discussion 1349-52; author reply 1352.

[19]   Madoff RD, Orrom WJ, Rothenberger DA, Goldberg SM. Rectal compliance: a critical reappraisal. Int J Colorectal Dis 1990;5:37-40.

[20]   Gregersen H, Djurhuus JC. Impedance planimetry: a new approach to biomechanical intestinal wall properties. Dig Dis 1991;9:332-40.

[21] Gregersen H, Andersen MB. Impedance measuring system for quantification of cross-sectional area in the gastrointestinal tract. Med Biol Eng Comput 1991;29:108-10.

[22] Bell AM, Pemberton JH, Hanson RB, Zinsmeister AR. Variations in muscle tone of the human rectum: recordings with an electromechanical barostat. Am J Physiol 1991;260:G17-25.

[23] Previnaire JG, Soler JM, Perrigot M, Boileau G, Delahaye H, Schumacker P et al. Short-term effect of pudendal nerve electrical stimulation on detrusor hyperreflexia in spinal cord injury patients: importance of current strength. Paraplegia 1996;34:95-9.

[24] Krogh K, Mosdal C, Gregersen H, Laurberg S. Rectal wall properties in patients with acute and chronic spinal cord lesions. Dis Colon Rectum 2002;45:641-9.

[25] Rasmussen OO, Ronholt C, Alstrup N, Christiansen J. Anorectal pressure gradient and rectal compliance in fecal incontinence. Int J Colorectal Dis 1998;13:157-9.

[26] Rao SS, Ozturk R, Stessman M. Investigation of the pathophysiology of fecal seepage. Am J Gastroenterol 2004;99:2204-9.

[27] Oettle GJ, Heaton KW. "Rectal dissatisfaction" in the irritable bowel syndrome. A manometric and radiological study. Int J Colorectal Dis 1986;1:183-5.

[28] Sorensen M, Rasmussen OO, Tetzschner T, Christiansen J. Physiological variation in rectal compliance. Br J Surg 1992;79:1106-8.

[29] Akervall S, Fasth S, Nordgren S, Oresland T, Hulten L. Rectal reservoir and sensory function studied by graded isobaric distension in normal man. Gut 1989;30:496-502.

[30] Kendall GP, Thompson DG, Day SJ, Lennard-Jones JE. Inter- and intraindividual variation in pressure-volume relations of the rectum in normal subjects and patients with the irritable bowel syndrome. Gut 1990;31:1062-8.

[31] Michelsen HB, Worsøe J, Krogh K, Lundby L, Christensen P, Buntzen S et al. Rectal motility after sacral nerve stimulation for faecal incontinence. Neurogastroenterology and Motility 2010;22:36,41+e6.

[32] Worsoe J, Michelsen HB, Buntzen S, Laurberg S, Krogh K. Rectal motility in patients with idiopathic fecal incontinence: a study with impedance planimetry. Dis Colon Rectum 2010;53:1308-14.

[33] Petersen P, Gao C, Rossel P, Qvist P, Arendt-Nielsen L, Gregersen H et al. Sensory and biomechanical responses to distension of the normal human rectum and sigmoid colon. Digestion 2001;64:191-9.

[34] Hammer HF, Phillips SF, Camilleri M, Hanson RB. Rectal tone, distensibility, and perception: reproducibility and response to different distensions. Am J Physiol 1998;274:G584-90.

[35] Andersen IS, Michelsen HB, Krogh K, Buntzen S, Laurberg S. Impedance Planimetric Description of Normal Rectoanal Motility in Humans. Dis Colon Rectum 2007.

Chapter   4.

Acute effect of electrical stimulation of the dorsal genital nerve on rectal capacity in patients with spinal cord injury

J. Worsøe[*#], *L. Fynne*[+], *S. Laurberg*[*], *K. Krogh*[+], *N.J.M. Rijkhoff*[#]

[#]*Center for Sensory-Motor Interaction (SMI), Department of Health Science and Technology, Aalborg University, Denmark*
[*]*Department of Surgery P, Aarhus University Hospital, Aarhus, Denmark*
[+]*Neurogastroenterology Unit, Department of Hepatology and Gastroenterology, Aarhus University Hospital, Aarhus, Denmark*

4.1 INTRODUCTION

Most subjects with spinal cord injury (SCI) have constipation and fecal incontinence, often resulting in restricted social activities and impaired quality of life [1]. Symptoms may be caused by abnormal rectal compliance and contractility, reduced anorectal sensibility, lack of external anal sphincter control and abnormal colorectal motility. The severity of neurogenic bowel dysfunction mainly depends on the completeness and level of injury but time since injury is important too [2,3]. Most authors have found that rectal compliance is reduced in patients with supraconal SCI and data suggests that it is increased in those with conal or cauda equina lesions [4,5]. One group has found increased rectal compliance in patients with conal or cauda equina lesions [6].

Neurogenic bowel dysfunction is usually treated conservatively with oral laxatives, suppositories and digital anorectal stimulation. Further treatment includes transanal irrigation, antegrade irrigation through an appendicostomy, colostomy, or electrical stimulation with the Brindley anterior root stimulator for assisted defecation [7,8].

Treatment is often unsatisfactory and new modalities should be explored. Stimulation of the dorsal genital nerve (DGN) can suppress vesical detrusor contractions and increase bladder capacity in patients with supraconal SCI [9]. Also, a pilot study has indicated that DGN stimulation can increase rectal compliance in SCI patients [10]. If data can be reproduced with other methods, DGN may have a future role in alleviating bowel symptoms in individuals with supraconal SCI.

The aim of the present study was to investigate whether DGN stimulation has an acute effect on the rectal CSA in patients with supraconal SCI.

4.2 METHODS

4.2.1   Subjects

Seven subjects with supraconal SCI were included (one female, median age 50 years; range: 39-67 years), median time since injury 19 years (range: 12-33 years). The lesion was motor and sensory complete in all the patients. Median neurogenic bowel dysfunction score was 14 (range: 5-19) [11]. Further demographics are given in table 1. The study was performed in accordance with the Helsinki Declaration II and was approved by the local ethical committee (M-20090145). All participants gave their fully informed written consent.

N	Gender	Age (years)	Time since injury (years)	Level of injury	ASIA score	NBD score
1	M	39	18	Th6	A	11
2	M	48	29	Th8	A	14
3	F	50	33	Th8	A	16
4	M	67	19	Th3	A	5
5	M	61	19	Th9	A	15
6	M	51	23	Th2	A	19
7	M	41	12	Th7	A	10

Table 4.1: Patient demographics. ASIA score: American spinal injury association score. ASIA score A indicate motor and sensory complete lesion. NBD: neurogenic bowel dysfunction. NBD score <9: minor dysfunction, 10-13: moderate dysfunction, ≥14: severe dysfunction.

4.2.2   Stimulation

Stimulation was performed using a constant current stimulator (Digitimer model DS7A, Digitme Ltd., Welwyn garden city, England) with the frequency controlled by a waveform generator (Hewlet-Packard model 33120A, USA). Square pulses with a pulse width of 200 μs and a frequency of 20 Hz were used. The amplitude was set at two times the threshold of the genito-anal reflex. One electrode (dimension: 10 x 20 mm, Neuroline 700, Ambu, Ballerup, Denmark) was placed at the base of the penis or on the clitoris as cathode, and a second electrode (diameter: 32 mm, PALS Platinium, Axelgaard, Lystrup, Denmark) was placed 2-3 cm lateral to the base of the penis or labia major. Precautions were taken to ensure good contact between skin and the electrodes. The genito-anal reflex threshold was identified visually during slowly stepwise increase of the amplitude before the investigation started.

4.2.3   Impedance planimetry

Rectal impedance planimetry allows simultaneous monitoring of rectal cross sectional area (CSA) and rectal pressure. The method avoids most sources of error associated with volume based methods [12]. At a constant current I, the potential difference (ΔV) between two detection electrodes and the CSA are proportionally related (CSA = I d σ-1 ΔV -1). The electrodes were 3 mm (d) apart and contained in a fluid with the conductivity σ. The rectal probe, used in the present study, had two excitation electrodes (60 mm apart), providing a sinusoidal current of 0.1 mA at 10 kHz, and a pair of detection electrodes (Fig. 2.1). The rectal CSA was measured approximately 60 mm above the anal verge. The electrodes were within a non-compliant flaccid bag (diameter: 90 mm, length: 90 mm, maximum volume: ≈570 ml), which was filled with 0.9 % saline at 37° C. The pressure within the bag was controlled by elevation of an open water container. Before measurements a multipoint calibration was done using circular plastic tubes with an inner diameter in the range from 283 mm2 to 4322 mm2. Intraluminal rectal pressure and anal pressure were measured with pressure transducers (Baxter, Deerfield, IL, USA) connected to perfused catheters within the bag and a within a balloon placed in the anal canal respectively. The pressure transducers were calibrated using 0 cmH2O and 100 cmH2O as minimum and maximum. All signals were sampled at 10 Hz and data were visualized and stored using custom made software (Openlab, Gatehouse, Noerresundby, Denmark). Before placement of the probe, resting rectal pressure was measured using a water-perfused catheter placed in the rectum. During the investigation, the patient was in the left lateral position and was not allowed to talk. The equipment for impedance planimetry has been described previously [13].

4.2.4   Distension protocol

A pressure controlled phasic distension protocol with a total length of 96 minutes was executed (Fig. 4.1). Before initiation of the phasic distensions, a distension pressure of 10 cmH2O above resting rectal pressure, lasting 12 minutes, was applied to condition the rectal wall. This was followed by four minutes distension

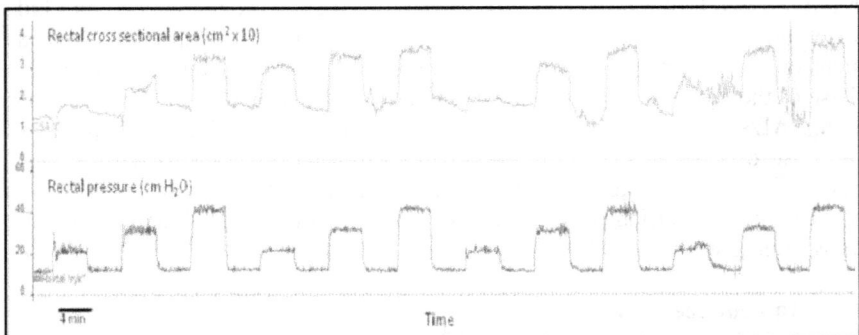

Figure 4.1: An example of a recording (patient #7). Four distension series with distension 10, 20, and 30 cmH2O above resting rectal pressure are shown. During the first 20 cmH2O distension, CSA increased suddenly in the end so therefore mean CSA was calculated during the steady state period before the last minute. DGN stimulation was done during the first and the third series.

at resting rectal pressure. Hereafter, distensions at 3 pressure levels (10, 20, and 30 cmH2O) were done. Each of the distensions lasted for four minutes and was separated by four minutes with the pressure at resting rectal pressure. This distension sequence was repeated four times, with DGN stimulation during 1st and 3rd distension series or 2nd and 4th distension series as randomized. Between the distensions, no DGN stimulation was applied.

4.2.5   Analysis

The CSA and rectal pressure were determined when steady state was present at each pressure level. Steady state CSA, calculated as the mean CSA for a one minute period, was assumed when the difference in mean CSA, for two 10 s periods one minute apart, was less than 10 percent (Fig. 4.2). Data that did not meet this requirement were discarded. For each distension pressure, the mean CSA from two distensions during stimulation was compared with the mean CSA from two distensions without stimulation. The rectal pressure-CSA relation (CSA/PR) was calculated for each pressure level. The circumferential wall tension was calculated using Laplaces' law $T = \Delta p\ r$, where T is the circumferential wall tension, r is the radius of the bag and Δp is the transmural pressure calculated as the difference between resting rectal pressure and rectal pressure during distension.

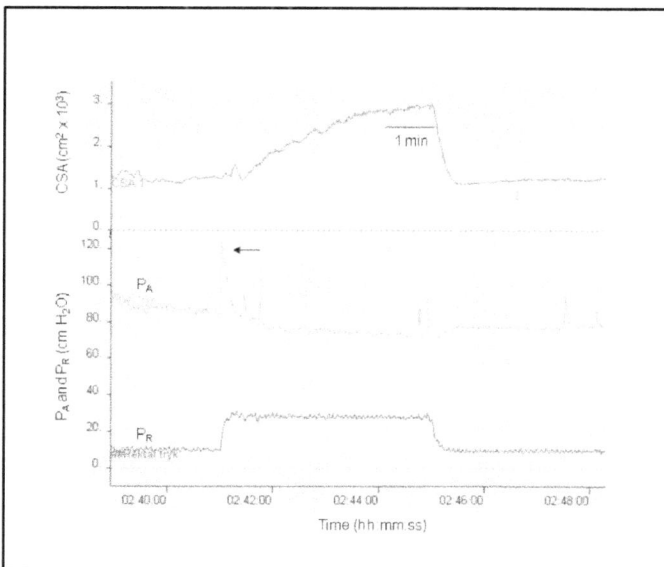

Figure 4.2: An example of a stimulated distension. A pressure controlled (PR) rectal distension is performed. When distension is started the stimulation is turned on, the anal sphincter contracts, which is seen as an increase in anal pressure (PA) (arrow). The rectal cross sectional area (CSA) increases and begins to stabilize after approximately two minutes.

4.2.6   Statistics

Numerical data are gives as medians with ranges. Statistical comparisons were made using Wilcoxon's test for non-parametric comparison of paired measurements.

4.3 RESULTS

All patients tolerated the investigation well. Three patients had lesions above Th6 and none of them experienced symptoms of autonomic dysreflexia during electrical stimulation. The median resting rectal pressure was 9.5 cmH2O (range: 8-16 cmH2O). Electrical stimulation above the reflex threshold could be seen as a brief (1-3 s) increase in anal pressure (Fig. 4.2). The median stimulation amplitude was 51 mA (range: 30-64 mA). Filling of the bag resulted in an increase in rectal CSA. Most of the increase occurred within the first 30 s, and thereafter, the rectal CSA became stable, with changes of less than 10% during the fourth minute in all of the patients. Data from the 10 cmH2O distensions were discarded since they were not reproducible and reliable.

The median CSA was smaller with than without stimulation in all 7 patients at 20 cmH2O distension pressure (p=0.02), and in 6 of 7 patients at 30 cmH2O distension pressure (p=0.03) (Fig. 4.3, Table 4.2). The median decrease in rectal CSA was 9% (7 cm2) at 20 cmH2O distension pressures and 4% (1 cm2) at 30 cmH2O at distension pressures. The rectal pressure-CSA relation was also significantly smaller during stimulation at 20 cmH2O (medians 1.0 cm2/ cmH2O vs. 1.1 cm2/ cmH2O) (p=0.03)) and 30 cmH2O distension (medians 0.9 cm2/ cmH2O vs. 0.9 cm2/ cmH2O) (p=0.02) (Table 3).The rectal wall tension was unchanged during stimulation (Table 4.3).

ID	20 cmH$_2$O			30 cmH$_2$O		
	Control CSA (cm^2)	Stimulation CSA (cm^2)	Relative change (%)	Control CSA (cm^2)	Stimulation CSA (cm^2)	Relative change (%)
1	44	40	-9	46	44	-4
2	26	20	-23	31	30	-3
3	32	28	-13	36	37	3
4	25	23	-8	29	26	-10
5	35	34	-3	38	36	-5
6	50	48	-4	55	53	-4
7	35	26	-26	37	35	-5
Median	35	28	-9	37	36	-4

Table 4.2: Rectal CSA during rectal distensions. Four consecutive distension series (20, and 30 cmH2O) were performed with stimulation during the 1st and 3rd or during the 2nd and 4th series as randomised. For each distension pressure, the mean value of the two distensions performed during stimulation and the two distensions performed without stimulation (control) was calculated. Mean values during and without stimulation was compared. CSA: cross sectional

Stimulation for Faecal Incontinence

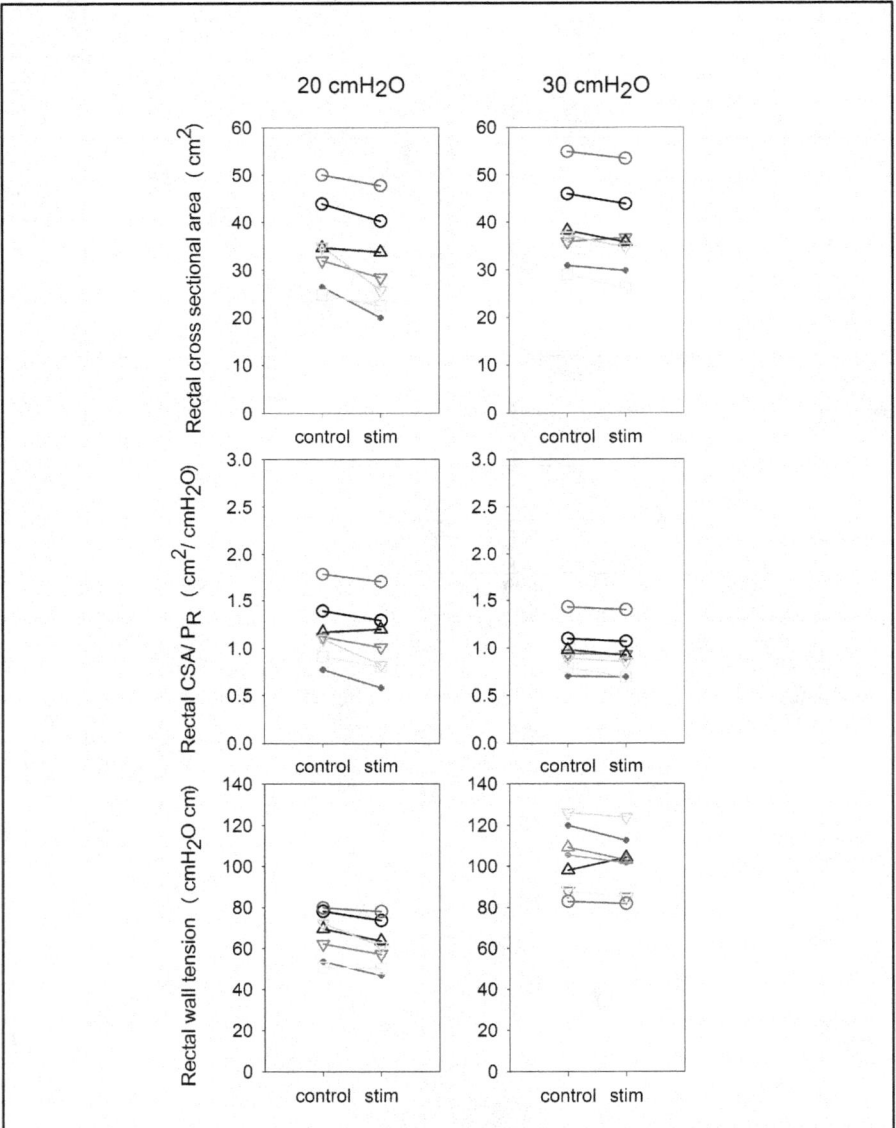

Figure 4.3: Data from all seven patients are shown. Stimulated (stim) and unstimulated (control) distensions are compared at different distension pressures (20 and 30 cmH_2O). Rectal CSA, the rectal CSA-pressure relation (CSA/PR), and rectal wall tension are shown.

	Distension pressure (cmH$_2$O)	Stimulation	Control	P-value
Median rectal CSA/ P$_R$	20	1.0 (0.6-1.7)	1.1 (0.8-1.8)	0.03
(cm^2 /cmH$_2$O)	30	0.9 (0.7-1.4)	0.9 (0.7-1.4)	0.02
Median rectal wall tension	20	61 (47-78)	70 (51-80)	0.02
(cmH$_2$O cm)	30	103 (82-124)	106 (83-126)	0.15

Table 4.3: The rectal CSA/PR, and rectal wall tension during the rectal distensions given as medians (range). PR: rectal pressure.

4.4 DISCUSSION

This study shows that the rectal CSA is reduced during acute DGN stimulation in subjects with complete supraconal SCI. The stimulation parameters were chosen based on experience with DGN stimulation to achieve inhibition of bladder contractions. A pulse width of 200 µs, a pulse rate of 20 Hz and an amplitude of at least twice the genito-anal reflex threshold have been used [9].

It has been demonstrated that the configuration of the distension profile (i.e. phasic, ramp or staircase) has only little effect on distensibility [14]. Furthermore, a randomized stimulation protocol was used to avoid bias from a potential carry over effect from stimulation and relaxation as a result of repeated distensions.

The relationship between pressure and CSA can be described by a first order system. The continuous decrease in CSA change per time unit inferred that longer distensions would produce more stable CSAs. This had to be balanced against the potential impairment of mucosa blood flow during prolonged distension and efforts to minimize the discomfort of the patients. Distensions lasting four minutes were chosen as a safe compromise. Preferably, calculation of mean CSA was done for the last minute of each distension. However during some distensions, changes in CSA larger than 10% were seen during the last minute. If a steady state period was present before the last minute, this was used to calculate the mean CSA.

Previously, we have used impedance planimetry for description of rectal CSA in SCI patients with supraconal SCI and reported a median rectal CSA of 11 cm2 during distension at 10 cmH2O and 18 cm2 during distension at 30 cmH2O [4].

Larger CSAs were measured in the present study. The patients from the two studies are, however, not directly comparable as in the previous study, patients were investigated just after injury and again one year later, while the median time since injury was 19 years in the present study. It has been shown, that constipation becomes more severe with time since injury and it is likely that the rectal wall properties change too [3].

During rectal distensions at 20 and 30 cmH2O, the median pressure-CSA relation was smaller during stimulation compared with the control distensions. Our findings are in contrast with data presented by Chung et al. who found increased rectal compliance during acute DGN stimulation [10]. They also used a stimulation amplitude of twice the reflex threshold but rectal compliance was measured using a barostat. Furthermore, they tested different stimulation frequencies (0.2, 2 and 20 Hz) and rectal compliance was larger during stimulation at 20 Hz compared to stimulation at a lower frequency. Chung et al. described a maximum increase of rectal compliance of 50% (at a rectal volume of 200 ml) during stimulation at 20 Hz [10]. The maximum rectal compliance was approximately 7 ml/ cmH2O without stimulation, and 12 ml/ cmH2O during stimulation. In that study, all the patients had complete supraconal lesions, but the duration of the injury was not mentioned. Rectal compliance both during and without stimulation were within the range of normal rectal compliance reported in the literature, which ranges from 4.5 ml/ cmH2O to 17 ml/ cmH2O [15,16]. This wide range of measured rectal compliance warrants cautiousness when comparing data between different centers.

Traditionally, rectal compliance is studied with pressure-volume based methods using balloons (ie barostat). The effect of rhizotomy and the response to feeding have been investigated in SCI patients [5,17]. There are however some methodological problems with these techniques [12]. Impedance planimetry determines rectal CSA, thereby avoiding some of the inherent sources of error with pressure-volume measurements [18]. The impedance planimetry probe used for this experiment was validated in vitro and accuracy was fair with a mean error of 7.3% (range: 0-14%). No low pass filter was included in the signal conditioning system (no anti-aliasing). However, it is unlikely that this had any influence on the results. At low distension pressures (10 cmH2O) the quality of rectal CSA measurements was not reliable. This could be caused by folding of the bag in irregular shape or eccentric position of the probe in the rectum. In another study ultrasound was used to confirm the correct positioning of the probe during distension [19].

Various implanted devices applying electrical stimulation have been used for treating neurogenic fecal incontinence. The use of Interstim Therapy has been investigated by Schurch et al., who performed a test stimulation in three SCI patients with complete lesions [20]. Both an early latency reflex corresponding to the genito-anal reflex and a late latency reflex, with high variability in latency were found indicating a polysegmental response. In none of the patients did the test stimulation reduce neurogenic incontinence suggesting that spino-bulbo-spinal pathways are necessary for sacral neuromodulation to be effective.

A previous study on the bladder showed inhibitory effects from DGN stimulation including suppression of bladder contractions, higher bladder capacity, and lower storage pressure [9]. Though we did not investigate phasic rectal contractions similar inhibitory effects could not be shown in this study. A fundamental difference between the bladder and the rectum is that the latter is modulated by the enteric nervous system, which could explain why results from stimulation of the bladder are not directly applicable to the bowel. It was hypothesized that DGN reduces neurogenic faecal incontinence by reducing rectal tone and contractility. This is not supported by the present study. Even though a reduction in CSA during stimulation was seen in all patients, the changes were relatively small. If this will have clinical implications remain to be studied. An alternative mode of action could be that DGN, by increasing rectal tone improves rectal emptying at defecation thereby reducing fecal incontinence. Larger studies of the effects of DGN stimulation are needed and it is possible that chronic effects may differ from those found in acute experiments.

4.5 CONCLUSION

In conclusion and in contrary to our hypothesis, it was shown that acute DGN stimulation in subjects with supraconal SCI results in reduced rectal compliance CSA.

REFERENCES

[1]   Glickman S, Kamm MA. Bowel dysfunction in spinal-cord-injury patients. Lancet 1996;347:1651-3.

[2]   Liu CW, Huang CC, Chen CH, Yang YH, Chen TW, Huang MH. Prediction of severe neurogenic bowel dysfunction in persons with spinal cord injury. Spinal Cord 2010.

[3]   Faaborg PM, Christensen P, Finnerup N, Laurberg S, Krogh K. The pattern of colorectal dysfunction changes with time since spinal cord injury. Spinal Cord 2008;46:234-8.

[4]   Krogh K, Mosdal C, Gregersen H, Laurberg S. Rectal wall properties in patients with acute and chronic spinal cord lesions. Dis Colon Rectum 2002;45:641-9.

[5]   Sun WM, MacDonagh R, Forster D, Thomas DG, Smallwood R, Read NW. Anorectal function in patients with complete spinal transection before and after sacral posterior rhizotomy. Gastroenterology 1995;108:990-8.

[6]   Storrie JB, Harding S, Raeburn AJ, Trivedi PM, Preziosi G, Emmanuel AV. Medium-term outcome with transanal irrigation for neurogenic bowel dysfunction is related to rectal compliance. Gut 2009;58:A87.

[7]   Christensen P, Bazzocchi G, Coggrave M, Abel R, Hultling C, Krogh K et al. A randomized, controlled trial of transanal irrigation versus conservative bowel management in spinal cord-injured patients. Gastroenterology 2006;131:738-47.

[8]   Furlan JC, Urbach DR, Fehlings MG. Optimal treatment for severe neurogenic bowel dysfunction after chronic spinal cord injury: a decision analysis. Br J Surg 2007;94:1139-50.

[9]   Hansen J, Media S, Nohr M, Biering-Sorensen F, Sinkjaer T, Rijkhoff NJ. Treatment of neurogenic detrusor overactivity in spinal cord injured patients by conditional electrical stimulation. J Urol 2005;173:2035-9.

[10]   Chung EAL, Woodhouse JB, Balasubramaniam V, Emmanuel AV, Craggs MD. Does sacral afferent stimulation influence bowel compliance? Neurourol Urodyn 2005;24:498-9.

[11]   Krogh K, Christensen P, Sabroe S, Laurberg S. Neurogenic bowel dysfunction score. Spinal Cord 2006;44:625-31.

[12]   Madoff RD, Orrom WJ, Rothenberger DA, Goldberg SM. Rectal compliance: a critical reappraisal. Int J Colorectal Dis 1990;5:37-40.

[13]   Gregersen H, Djurhuus JC. Impedance planimetry: a new approach to biomechanical intestinal wall properties. Dig Dis 1991;9:332-40.

[14]   Hammer HF, Phillips SF, Camilleri M, Hanson RB. Rectal tone, distensibility, and perception: reproducibility and response to different distensions. Am J Physiol 1998;274:G584-90.

[15]   Oettle GJ, Heaton KW. "Rectal dissatisfaction" in the irritable bowel syndrome. A manometric and radiological study. Int J Colorectal Dis 1986;1:183-5.

[16]   Sorensen M, Rasmussen OO, Tetzschner T, Christiansen J. Physiological variation in rectal compliance. Br J Surg 1992;79:1106-8.

[17]   Suttor VP, Ng C, Rutkowski S, Hansen RD, Kellow JE, Malcolm A. Colorectal responses to distension and feeding in patients with spinal cord injury. Am J Physiol Gastrointest Liver Physiol 2009;296:G1344-9.

[18]   Krogh K, Ryhammer AM, Lundby L, Gregersen H, Laurberg TS. Comparison of methods used for measurement of rectal compliance. Dis Colon Rectum 2001;44:199-206.

[19] Andersen IS, Michelsen HB, Krogh K, Buntzen S, Laurberg S. Impedance Planimetric Description of Normal Rectoanal Motility in Humans. Dis Colon Rectum 2007.

[20] Schurch B, Reilly I, Reitz A, Curt A. Electrophysiological recordings during the peripheral nerve evaluation (PNE) test in complete spinal cord injury patients. World J Urol 2003;20:319-22.

Chapter   5.

Gastric transit and small intestinal transit time and motility assessed by a magnet tracking system

J. Worsøe[*#], *L. Fynne*[□], *T. Gregersen*[□], *V. Schlageter*[+], *L.A. Christensen*[++], *J.F. Dahlerup*[++], *N.J.M. Rijkhoff*[#], *S. Laurberg*[*], *K. Krogh*[□]

[*]*Department of Surgery P, Aarhus University Hospital, Aarhus, Denmark*
[#]*Center for Sensory-Motor Interaction (SMI), Department of Health Science and Technology, Aalborg University, Denmark*
[□]*Neurogastroenterology Unit, Department of Hepatology and Gastroenterology V, Aarhus University Hospital, Aarhus, Denmark*
[+] *Motilis Medica SA, Lausanne, Switzerland*
[++]*Department of Hepatology and Gastroenterology V, Aarhus University Hospital, Aarhus, Denmark*

5.1 INTRODUCTION

The prevalence of gastrointestinal motility and functional gastrointestinal disorders is high in the general population [1,2]. Furthermore symptoms of disturbed GI motility are often a significant problem in patients with other medical problems. Diagnosing and alleviating these disorders require good evaluation methods that can identify abnormal GI physiology. Gastrointestinal motility is usually described in terms of regional transit times or as intraluminal pressure changes. Scintigraphy is the gold standard for determination of gastric emptying and small intestinal transit [3,4]. Contraction patterns have been investigated using manometry catheters. Solid state catheters with small pressure transducers have facilitated ambulatory examinations and allowed recording of diurnal variation [5-7]. Disadvantages of these techniques include the invasiveness, the exposure to radiation and that they are relatively expensive. The hydrogen breath test is an alternative for determination of transit times, but it is affected by small intestinal bacterial overgrowth and does not distinguish between gastric and intestinal transit times [8].

New techniques aim to improve the quality of motility data and also to reduce the side effects and patient discomfort. Video capsule endoscopy, primarily used for evaluation of the small intestinal mucosa pathology, may be an alternative for determination of transit times [9,10]. However, for the mere purpose of obtaining transit times, it is expensive and analysis is time consuming. Computerized picture analysis of capsule endoscopy images has recently been used for description of small intestinal motility patterns [11]. Lately, a wireless motility capsule (Smartpill) that measures temperature, pressure and pH has been used to investigate segmental and whole-gut transits [12,13]. Magnetic resonance imaging have also been used to measure gastric and small intestinal motility [14,15]. MRI has also been used to track the position of fluorine labeled capsules giving information about small intestinal motility patterns and this can be combined with anatomical data [16].

Information about motility patterns and transit can also be obtained by tracking a small magnet through the gastrointestinal tract. Early methods based on ingestion of a small solid magnet have been refined to improve spatial and temporal resolutions [17-21]. High resolution data on gastrointestinal transit have been obtained using multi-channel superconducting quantum interference, but the equipment is expensive and requires a shielded environment [22-24]. Magnetic moment imaging using a tracking system with anisotropic magneto-resistor sensors was recently validated with scintigraphy demonstrating good correlation between gastric transit time and positional data [25]. The Magnet Tracking System (MTS-1; Motilis, Lausanne, Switzerland) has been developed for use in a standard laboratory setting [26,27]. MTS-1 has been used in animal studies, studies in healthy humans, and in patients with neurogenic bowel dysfunction [28-33]. However, a validation with simultaneous measurements using established methods is needed. If the difference in contraction frequency and position of the magnet measured with MTS-1 can be used to determine pyloric and ileocecal passages, then MTS-1 will be an easy, minimally-invasive, and non-radiant tool to provide valid information on gastric transit time and small intestinal transit times.

The primary aim of this study was to investigate if MTS-1 could be used to reliably determine gastric transit and small intestinal transit time. Data from simultaneous capsule endoscopy was used as reference. Furthermore, small intestinal motility patterns recorded with MTS-1 in the fasting state and in the postprandial state were compared for identification of migrating motor complex phase III during fast. An algorithm was applied for classification of fast movements, slow movements, and very slow movements and by comparing small intestinal contraction frequencies.

5.2   METHODS

5.2.1   subjects

Eight healthy volunteers (3 males, median age 30 years; range: 25-61 years), with median BMI 21.3 kg m-2 (range: 20.2-26.5 kg m-2) were included. No subjects had undergone abdominal surgery, were taking medication or suffered from diseases affecting gastrointestinal motility. All participants signed informed written consent and the study was approved by the local scientific ethical committee (M-20080037).

5.2.2   Magnet Tracking System, MTS-1

Subjects ingested a small magnetic pill (dimensions: 6x15 mm, weight: 0.9 g, density: 1.8 g cm-3, magnetic moment 0.2 Am2), which was tracked by a matrix of 4x4 magnetic field sensors separated by 5 cm and placed over the abdomen.The position of the sensor matrix with respect to anatomical landmarks was noted (iliac spines, intercostal angle, pubic bone) (Fig. 5.1). With a sampling rate of 10 Hz, each sensor measured the magnetic induction dependent on the distance between the sensors and the magnetic pill and the orientation of the pill. The position and orientation of the magnetic pill was defined by 5 coordinates (position: x, y, z, and angle: θ, φ).The change in position coordinates reflected propagation of the magnet. The change of the angles reflected change in orientation, which correlated with the contraction frequency of the relevant gastrointestinal segment. Data were analysed on a computer running custom-made software (MTS_Record, Motilis, Lausanne, Switzerland) showing real-time position and orientation of the magnetic pill (Fig. 5.1). Before the measurements began, the sensors were calibrated by offsetting the earth's magnetic field. Artefacts due to respiration and movement were recorded using accelerometers placed on the abdomen and the neck. During post processing, an adaptive algorithm was used to filter out movements in phase with the respiration.

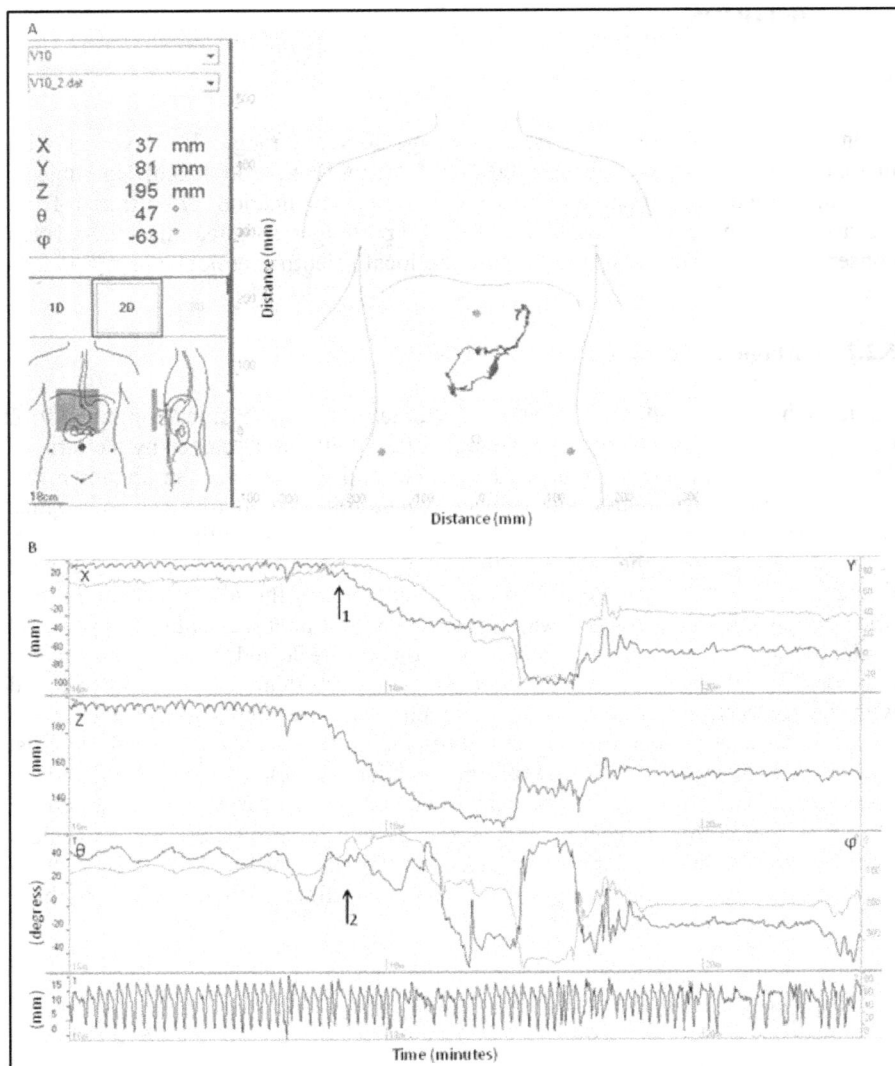

Figure 5.1: Real time recording with the MTS-1 (example from one subject). 1A: The position x, y and z and orientation θ and φ are displayed. Position of the sensor array over the body is seen to the left. To the right a recording of the movement through the duodenal arch is displayed. 1B: Duodenal passage (from 17m40s to 19m30s) is seen as a change in position (x, y and z) (arrow 1) and disappearance of the characteristic 3 contractions min-1 pattern of the stomach (θ and φ) (arrow 2). The curve at the bottom shows noise from respiration and movement.

5.2.3   MTS-1 combined with PillCam

The validity of gastric transit and small intestinal transit determined with MTS-1 was tested through comparison with the simultaneous use of a PillCam (Fig. 5.2). The video capsule (PillCam, Given, Yoqnaem, Israel) measures 11x26 mm and contains an imaging device (field of view of 156°) and a light source at one end of the capsule [34]. Images were transmitted at a rate of two images s-1 with a battery powered light source lasting for a minimum of eight hours.

A magnet-PillCam unit was constructed by gluing (Loctite 4013 medical line, Henkel, Rocky Hill, CT, USA) the magnetic pill on a PillCam and covering the magnet-PillCam unit with a polyurethane sheet.

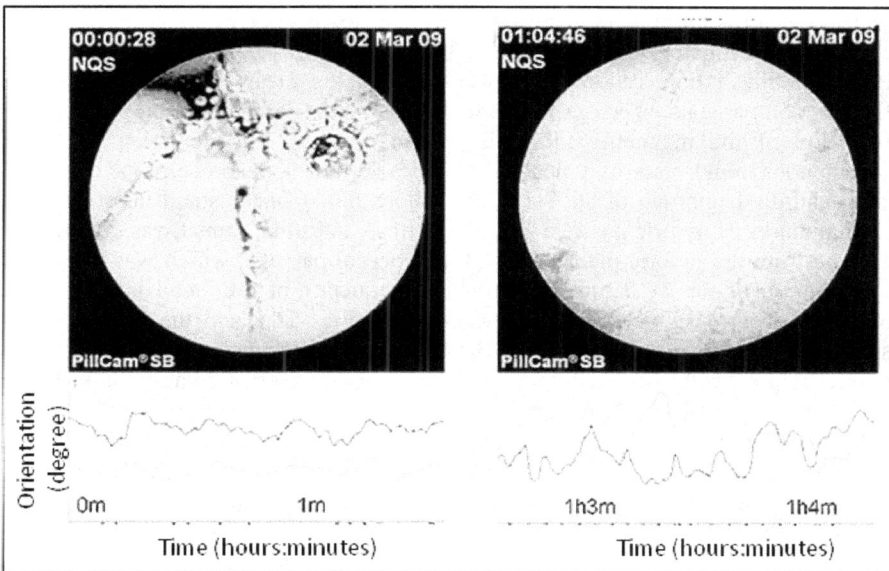

Figure 5.2: Correlation of anatomical data and motility data using the PillCam and the Magnet Tracking System , Left: image from the stomach with simultaneous MTS-1 data (orientation θ and φ on the y-axis) showing a contraction frequency of approximately 3 min-1 (arbitrary unit), consistent with localization in the stomach. Right: image from the proximal small intestine with simultaneous MTS-1 data showing a contraction frequency of approximately 9-10 min-1 (arbitrary unit) consistent with localization in the small intestine.

5.2.4   Protocol

The subjects underwent three experiments on three separate days, all starting at 9 AM and continuing for six to eight hours with the following steps: (1) ingestion of the magnet-PillCam unit where a standard meal (≈1500 kJ, 32% fat) was given after four hours investigation in the fasting state with the investigation continued until ileocecal passage; (2) ingestion of the magnetic pill alone in a similar setting as (1); and (3) ingestion of the magnetic pill followed by a standard meal given right after pyloric passage (≈ 2200 kJ, protein, 48% fat). During the investigations, subjects were placed in a bed with head elevation (>45°) or lying down. They were encouraged not to talk or move. The recordings were interrupted for small breaks on request.

5.2.5   Analysis

Experiment (1) was used to test the validity of MTS-1 for assessment of gastric transit and small intestinal transit time. Experiments (1) and (2) were used to compare gastric transit and small intestinal transit of two different sized objects. Experiments (2) and (3) were used to compare the fasting and the postprandial motility patterns for two hours after pyloric passage.

Two investigators independently determined the gastric transit and the small intestinal transit time in each investigation, and the mean times were used for further comparisons. The gastric transit time was defined as the time from ingestion of the magnetic pill until pyloric passage. The cessation of the 3 contractions min-1 pattern, typical for the stomach, the appearance of the duodenal arch, and the beginning of the 8-11 contractions min-1 of the small intestine were the hallmarks of pyloric passage (Fig. 5.1). Small intestinal transit was determined as time from the pyloric passage until the ileocecal passage, which was identified as cessation of the 8-10 min-1 contraction frequency of the small intestine(Fig. 5.3), the occurrence of a short fast movement (Fig. 5.4), and the magnetic pill situated in the lower right quadrant. The contraction frequencies were displayed in a time-frequency plot with a color code indicating the contraction amplitude. These data were obtained by computing the short-time Fourier transform (STFT) (Fig. 5.3). For this purpose, custom made software was used (MTS_Tool, Motilis, Lausanne, Switzerland). A standard approach for analysis of time-frequency maps was used. The power spectral density is estimated by and fast fourier transform on a short segment of data. A time frame of 3 min was used, and a Hamming window was applied. Calculations for the sliding window were conducted every 10 samples giving a time-frequency map. At each instant, peaks detection is applied to select main present frequencies. Only steady values were considered and extreme values were omitted based on Bayesian algorithms.

Capsule endoscopy with PillCam was used as the gold standard for detection of pyloric and ileocecal passage. Using PillCam images, gastric transit time was defined as time from ingestion of the magnet-PillCam unit until the time of the first picture in the duodenum. Small bowel transit was defined as the time from pyloric passage until the first picture of ileocecal passage. The PillCam recordings were examined by two experts and the mean value of their results was used as reference.

Motility patterns were analysed with Motilis-dedicated software for the upper gastro-intestinal tract (MTS_Tool, Motilis, Lausanne, Switzerland). The mean small intestinal propagation velocity for two hours following pyloric passage was computed. The mean contraction frequencies of the stomach and the small intestine were calculated using the contractions with the highest amplitudes obtained when the magnet was performing very slow movements (i.e. when there was no progression of the magnet). The mean contraction frequency in the small intestine was calculated using only contractions with a frequency higher than 6 min-1. The frequency peaks were selected using a convolution of the fast fourier transform with the "shape of a peak" described by a Gaussian function. The frequencies obtained during progression of the magnet were discarded, while frequencies obtained when the magnet did not progress were included. With this approach the Doppler effect (contraction frequency as a function of velocity of the magnet) was evaded.   A linear regression was used to derive the change in instantaneous contraction frequency during the first two hours after pyloric passage (Fig. 5.5).

An initial analysis of velocity histograms identified a trimodal distribution of velocities and the cut offs were made to separate the three types of movements velocities: fast movements (> 15 cm min-1), slow movements (between 1.5 and 15 cm min-1), and very slow movement (<1.5 cm min-1). Based on this analysis, an algorithm was developed for automatic classification of movements in the small intestine [29].

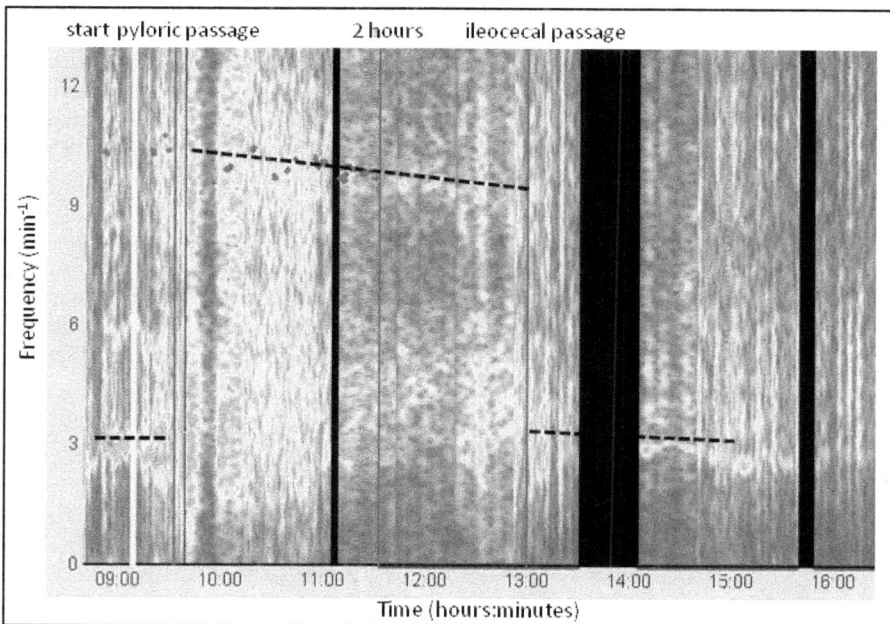

Figure 5.3: Time frequency plot. The contraction frequencies (dotted line) are illustrated as a function of time. A dominant frequency of 3 min-1 is seen initially as the magnet pill is located in the stomach. At approximately 09:45 the magnetic pill enters the small intestine and the dominant frequency changes to 10 min-1. Ileocecal passage is seen at approximately 13:00 as a drop in frequency to 4-5 min-1.

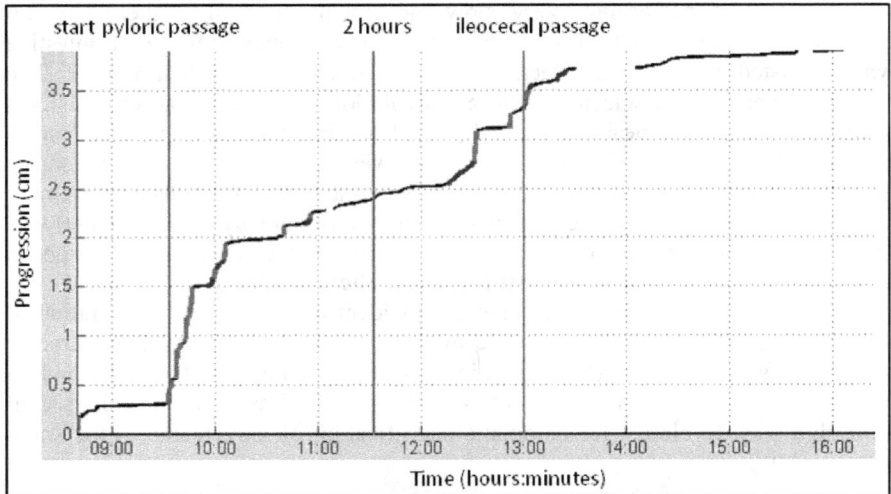

Figure 5.4: Progression of the magnetic pill over time during fast. Investigation made in fasting conditions. Pyloric passage, 2 hour after pyloric passage, and ileocecal passage is marked. Different velocities can be calculated (>15 cm min-1, <15 cm min-1, and <1.5 cm min-1). The contractions frequency can only be calculated when progression is very slow (<1.5 cm min-1). Most distance through the small intestine is covered during the period just after pyloric passage and during the period just before ileocecal passage. These two periods, separated by approximately 90 minutes, probably reflect phase III of the MMC.

5.2.6   Statistics

Numerical data are given as means and standard deviations and non-gaussian distributed data are given as medians and total range. Statistical significance was tested with Wilcoxon's test (non-parametric test for paired data), and the level of significance was set at 0.05.

5.3   RESULTS

5.3.1   Inter observer variation

For the MTS-1 investigations, the median difference between the two observers' determination of transit times was 1 min (range: 0-11 min) for gastric transit and 6 min (range: 0-33 min) for small intestinal transit.

5.3.2   Validation of gastric transit and small intestinal transit data determined with MTS-1

In all subjects, the magnet-PillCam unit was easily ingested and passed the cardia within 30 s. No pathology was seen in the stomach or in the small intestine. Pyloric passage was identified with the PillCam in all eight subjects. In one subject the magnet-PillCam unit underwent numerous regurgitations forth and back in the pyloric region before definitive pyloric passage. Agreement between gastric transit times determined with MTS-1 (median 56 min; range: 5-133 min) and with PillCam (median 57.5 min; range: 7-127 min) was good with a median difference of 1 min (range: 1-6 min) with no systematic difference (Table 5.1).

Ileocecal passage was identified using PillCam in seven subjects. In one subject, ileocecal passage could not be identified during the eight-hour investigation. Usually the magnet-PillCam unit was situated in the terminal ileum for a length of time (median 57 min; range: 19-148 min) before ileocecal passage. The small intestinal transit time determined with MTS-1 (median 255 min; range: 209-398 min) and PillCam (median 275 min; range: 209-398 min) showed good agreement as the median difference was 1 min (range: 0-52 min) with no systematic difference (Table 5.1).

| | Magnet-PillCam unit | | | | Magnetic pill alone | |
| | PillCam | | MTS-1 | | MTS-1 | |
Subject ID	Gastric transit (min)	Small intestinal transit (min)	Gastric transit (min)	Small intestinal transit (min)	Gastric transit (min)	Small intestinal transit (min)
1	127	-	133	-	73	402
2	29	241	30	241	53	251
3	19	292	20	284–294	4	260
4	60	307	60	255	52	-
5	7	275	5	276	48	292
6	55	209	53	209	17	261
7	107	245	107	245	23	-
8	60	398	59	398	18	241
Median	57.5	275	56	255	35.5	260.5

Table 5.1: Gastric transit and small intestinal transit times obtained using the magnet-PillCam unit and the magnetic pill in eight subjects. The magnet-PillCam unit was ingested during fast and a meal was given after four hours. In subject number four, ileocecal passage took place during a ten minutes break. Ileocecal passage determined with capsule endoscopy took place after eight minutes into the break, and error of 8 min was used for comparison with the PillCam. In three of sixteen investigations, ileocecal passage did not occur during the eight hour protocol.

5.3.3　Fasting and postprandial propagation velocities in the small intestine

The two-hour motility data during fasting and postprandially are presented in table 5.2. The median two-hour propagation velocity was 2.2 cm min-1 (range: 1.1-2.8 min) during the fasting, and 2.3 cm min-1 (range: 1.7-3.8 min) postprandially (p=0.50). Most small intestinal transit occurred through very fast movements (> 15 cm min-1) accounting for median 60% (range: 34-62%) of the distance in median 3% (range: 2-4%) of the time during the fast. Likewise in the postprandial state, 60 % (range: 42-74%) of the distance occurred with very fast movements in median 3% (range: 2-7%) of the time.

Fasting								
Subject ID	Fast movements (>15 cm min⁻¹)		Slow movements (<15 cm min⁻¹)		Very slow movements (<1.5 cm min⁻¹)		Mean contraction frequency (min⁻¹)	Mean progression velocity (cm min⁻¹)
	(cm)	(min)	(cm)	(min)	(cm)	(min)		
1	111	4	33	17	43	99	9.78	1.6
2	100	4	29	12	40	104	9.48	1.4
3	58	2	14	10	22	108	9.32	0.8
4	162	5	77	35	42	80	10.27	2.3
5	56	2	43	19	14	99	10.14	0.9
6	79	3	108	44	45	73	9.92	1.9
7	65	2	45	20	34	98	10.14	1.2
8	95	4	49	20	11	96	10.15	1.3
Median	87	3.5	44	19.5	37	98.5	9.90	1.4
Postprandial								
1	91	3	79	36	18	81	10.32	1.6
2	71	2	79	36	18	82	10.25	1.4
3	97	3	21	8	46	109	10.72	1.4
4	143	5	11	6	41	109	9.33	1.6
5	83	4	59	27	30	89	10.56	1.4
6	219	8	85	42	39	70	11.04	2.9
7	108	5	31	16	39	99	11.00	1.5
8	88	3	6	4	25	113	11.02	1.0
Median	94	3.5	45	21.5	34.5	94	10.53	1.5

Table 5.2: Fasting and postprandial motility for two hours after pyloric passage.Progression (cm) and duration (min) of fast (>15 cm min-1), slow (between 1.5 and 15 cm min-1), and very slow (<1.5 cm min-1) movements during fast and after a standard meal. The mean progression velocities during two hours are also given.

5.3.4   Transit and small intestinal motility patterns of magnetic pill versus magnet-PillCam unit

For the magnetic pill, median gastric transit time was 35.5 min (range: 4-73 min); median small intestinal transit time was 260.5 min (range: 241-402 min) (table 5.1). This finding did not differ significantly from transit times of the magnet-PillCam unit (p=0.21, p=0.89). There was no significant difference between the median two-hour propagation velocity with the magnetic pill (median 1.3 cm min-1; range: 0.8-2.3 min) and the larger magnet-PillCam unit (median 1.5 cm min-1; range: 1.0-1.7 min) (p=0.89). In one subject, there was a difference of 52 min between small intestinal transit determined with capsule endoscopy and MTS-1. In subject number four, ileocecal passage occured during a 10 min break. Ileocecal passage determined with capsule endoscopy occurred after an 8 min of the break, so a maximum error of 8 min was used for calculation. In two of the investigations with the magnetic pill and in one of the investigations with the magnet-PillCam unit, the ileocecal passage did not occur during the eight-hour investigation (Table 5.1).

5.3.5   Frequency of contractions

The mean contraction frequency of the stomach was 2.85 ± 0.29 min-1. Movements through the duodenum were fast (mean propagation velocity: 28 cm s-1 ± 20 cm s-1) and often separated by one or two pauses. The mean contraction frequencies in the small intestine was 9.90 ±0.14 min-1 for two hours during fast and 10.53 ±0.16 min-1 postprandially (p=0.03). The mean contraction frequency decreased during the first two hours after pyloric passage both during fasting and postprandially. Compared with postprandially (-1.12 min-1 cm-1), the slope during fasting was less step (-0.49 min-1 cm-1) (p=0.04) (Fig. 5-5).

5.4   DISCUSSION

The MTS-1 is a non-radiant and minimally invasive tool to determine gastrointestinal transit times. MTS-1 is accurate for determination of colorectal transit time, and pilot data on gastric and small intestine contraction patterns and transit times have been published [29,30]. In the present study we found that MTS-1 is valid for determination of gastric transit and small intestinal transit times. The inter-observer variation for assessment of gastric transit was low and not expected to be clinically relevant. Video capsule endoscopy was used as the "gold standard", and agreement between the two methods was good. Estimates of pyloric and ileocecal passage were based on the position of the magnet pill in the frontal plane and changes in the frequency spectrum as a function of time. The latter were recognizable and characteristic for the stomach, the small intestine, and the colon. An algorithm for analysing the time frequency plots may allow development of automatic determination of gastric transit and small intestinal transit time.

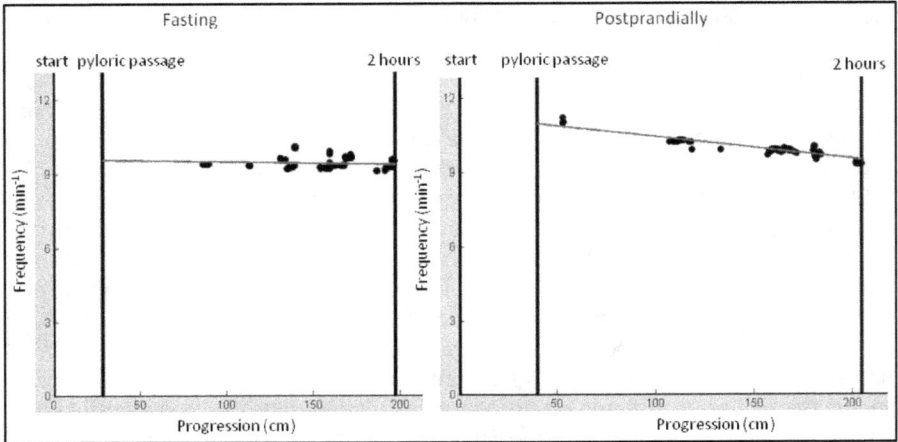

Figure 5.5: Small intestinal contraction frequency during fast and postprandially (example from investigations in one subject). The progression in the small intestine is on the x-axis, and the contraction frequency in the small intestine for two hours after pyloric passage is on the y-axis. In general, the mean contraction frequency was lower during fast compared to postprandially (9.48 min-1 vs. 10.25 min-1). The decrease in contraction frequency per 2 hours was smaller during fast when compared to postprandially (-0.18 Hz cm-1 vs. -1.45 Hz cm-1).

The precision of MTS-1 depends on the position and orientation of the magnet with respect to the sensor matrix. With only one sensor positioned 100 mm from the magnetic pill, the positioning error in the frontal plane is 10 mm, however changes in orientation of only 1-2 degrees can be detected [27]. This error is reduced by adding more sensors in a matrix, and the currently used system can track the magnetic pill at distances of more than 200 mm. The absolute accuracy of MTS-1 is approximately 1-2 cm, which is sufficient for anatomical localization. The amplitude of small back and forth movements can be measured more accurately (1-2 mm, rotation of 0.5°). With good quality recordings of the respiratory rhythm, the correction of respiratory artifacts at all amplitudes was accurate. A problem with the MTS-1 is that movement of the small intestine inside the abdomen affects the measurements. This can only be overcome with simultaneous collection of anatomical data (computer tomography), not included in this protocol. Thus, the distance covered and the velocity of the magnet pill reflects movement of the intestine as well as antegrade and retrograde intraluminal movement. The former is probably of minor importance but given the lack of distinction between back and forth mixing movements and short regular antegrade movements, velocity of the magnet pill should be considered a motility index rather than an estimate of progression through the intestine. However, most of the distance was covered during fast or very fast movement and those were clearly identifiable. A shortcoming of our protocol was that short breaks were allowed during the investigation, potentially influencing measurements of distance and calculation of total distance travelled in the small intestine. However, using the positioning of the sensor with respect to anatomical landmarks indicated that this error was very little.

Compared to scintigraphy, MTS-1 has no risk of radiation exposure; this is especially important if children are investigated. Scintigraphy, however, allows determination of gastric emptying for both solids and liquids (i.e. meals and macronutrients), whereas magnetic tracking only determines transit of the magnetic pill, since a small solid will leave the stomach with a phase III MMC [35]. Given the size of the magnet pill it is possible that its passage through the small intestine will differ from the passage of a meal. The same holds true for other methods including the wireless motility capsule (Smartpill) and the PillCam. Future comparison with scintigraphy may clarify this aspect. In the present study the meal was given to induce the postprandial small intestinal motility pattern when the magnet pill had reached the doudenum. The postprandial state affects the whole small intestine and we, therefore, consider the observed differences between fast and postprandial states valid even if the magnet pill did not behave entirely like the meal.

Capsule endoscopy has been used to assess small intestinal motility [11]. However, the size of the PillCam may affect contractions and transit [36]. Data from the present study seem to contradict this, as transit times with the specially constructed magnet-PillCam unit did not differ from those obtained with the much smaller magnetic pill.

Antroduodenal and small intestinal manometry is used clinically in the evaluation of patients with suspected severe dysmotility such as chronic intestinal pseudoobstruction [5,37]. It was anticipated that MTS-1 could be used for identification of phase III in the migrating motor complexes (MMC) and in the recordings during fasting we saw several examples of suggested MMC phase III (Fig. 4). However, no statistical difference in the distribution of fast movements which could represent phase III MMC was seen when comparing fasting and postprandial motility data. Future studies combining manometry and MTS are needed to validate changes in MMC seen by MTS. The propagated distance of the magnetic pill was the same during fasting and postprandially. During fasting the contraction frequency decreased in the aboral direction; this finding was even more pronounced postprandially, which likely reflects the small intestine adapting to intake of food and to promoting absorption. Similarly, the mean contraction frequency in the small intestine increased postprandially. A linear fitting was used to analyse contractions in the small intestine. It is recognised, that this model does not take into account the magnets progression velocity changes along the small intestine. Also, only data obtained when the magnet was performing very slow movements were included explaining why more data points exist at the end of the two-hour period. With further improvement of the analyses, it may become possible to identify motility patterns with pathological significance.

Recently, the wireless motility capsule (Smartpill, SmartPill Corporation, Buffalo, NY, USA) has been introduced. It is for ambulatory use, and measures pressure, pH, and temperature throughout the gastrointestinal tract [35,38]. The

Smartpill provides reliable information about gastric transit, small intestinal transit, total colonic transit, and some contraction patterns [12,13]. It is correct that most parameters obtained with MTS are also available with the SmartPill. Also, the SmartPill is developed into a clinically useful design which the MTS is not. There are two major differences: 1). SmartPill detects pressure whereas MTS detects movement. Measuring pressure in a moving object where one does not know the direction of the pressure sensor according to the lumen, direction of movement, or bowel wall involves major sources of error. These are avoided by MTS. 2). SmartPill determines total colorectal transit time, but allows no estimation of right versus left colonic transit. MTS tracks position and thereby potentially allows determination of segmental colonic transit. This may be clinically important.

In general, variations in gastric transit time are large and account for much of the variation in oro-cecal transit time [39]. Accordingly, we found large inter-subjective variations in gastric transit and small intestinal transit times. This finding is mainly caused by lack of timing of magnetic pill ingestion with phase III of MMC.

5.5   CONCLUSSION

In conclusion, the MTS-1 system is a promising minimally-invasive, non-radiant research tool that avoids radiation for the investigation of gastrointestinal motility. The method is accurate for the assessment of gastric transit and small intestinal transit times. Furthermore, it was possible to distinguish between fasting and postprandial small intestinal mean contraction frequencies, which warrants further exploration.

REFERENCES

[1]   Drossman DA, Li Z, Andruzzi E, Temple RD, Talley NJ, Thompson WG et al. U.S. householder survey of functional gastrointestinal disorders. Prevalence, sociodemography, and health impact. Dig Dis Sci 1993;38:1569-80.

[2]   Russo MW, Wei JT, Thiny MT, Gangarosa LM, Brown A, Ringel Y et al. Digestive and liver diseases statistics, 2004. Gastroenterology 2004;126:1448-53.

[3]   Edelbroek M, Horowitz M, Maddox A, Bellen J. Gastric emptying and intragastric distribution of oil in the presence of a liquid or a solid meal. J Nucl Med 1992;33:1283-90.

[4]   Argenyi EE, Soffer EE, Madsen MT, Berbaum KS, Walkner WO. Scintigraphic evaluation of small bowel transit in healthy subjects: inter- and intrasubject variability. Am J Gastroenterol 1995;90:938-42.

[5] Hansen MB. Small intestinal manometry. Physiol Res 2002;51:541-56.

[6] Rao SS, Sadeghi P, Beaty J, Kavlock R. Ambulatory 24-hour colonic manometry in slow-transit constipation. Am J Gastroenterol 2004;99:2405-16.

[7] Camilleri M, Bharucha AE, di Lorenzo C, Hasler WL, Prather CM, Rao SS et al. American Neurogastroenterology and Motility Society consensus statement on intraluminal measurement of gastrointestinal and colonic motility in clinical practice. Neurogastroenterol Motil 2008;20:1269-82.

[8] Camboni G, Basilisco G, Bezzani A, Bianchi PA. Repeatability of lactulose hydrogen breath test in subjects with normal or prolonged orocecal transit. Dig Dis Sci 1988;33:1525-7.

[9] Selby W. Complete small-bowel transit in patients undergoing capsule endoscopy: determining factors and improvement with metoclopramide. Gastrointest Endosc 2005;61:80-5.

[10] Tang SJ, Zanati S, Dubcenco E, Christodoulou D, Cirocco M, Kandel G et al. Capsule endoscopy regional transit abnormality: a sign of underlying small bowel pathology. Gastrointest Endosc 2003;58:598-602.

[11] Malagelada C, De Iorio F, Azpiroz F, Accarino A, Segui S, Radeva P et al. New insight into intestinal motor function via noninvasive endoluminal image analysis. Gastroenterology 2008;135:1155-62.

[12] Kuo B, McCallum RW, Koch KL, Sitrin MD, Wo JM, Chey WD et al. Comparison of gastric emptying of a nondigestible capsule to a radio-labelled meal in healthy and gastroparetic subjects. Aliment Pharmacol Ther 2008;27:186-9.

[13] Rao SS, Kuo B, McCallum RW, Chey WD, DiBaise JK, Hasler WL et al. Investigation of colonic and whole-gut transit with wireless motility capsule and radiopaque markers in constipation. Clin Gastroenterol Hepatol 2009;7:537-44.

[14] Schwizer W, Steingoetter A, Fox M. Magnetic resonance imaging for the assessment of gastrointestinal function. Scand J Gastroenterol 2006;41:1245-60.

[15] Froehlich JM, Patak MA, von Weymarn C, Juli CF, Zollikofer CL, Wentz KU. Small bowel motility assessment with magnetic resonance imaging. J Magn Reson Imaging 2005 21:370-5.

[16] Hahn T, Kozerke S, Schwizer W, Fried M, Boesiger P, Steingoetter A. Visualization and quantification of intestinal transit and motor function by real-time tracking of (19) F labeled capsules in humans. Magn Reson Med 2011.

[17]   Wenger MA, Henderson EB, Dinning JS. Magnetometer method for recording gastric motility. Science 1957;125:990-1.

[18]   Basile M, Neri M, Carriero A, Casciardi S, Comani S, Del Gratta C et al. Measurement of segmental transit through the gut in man. A novel approach by the biomagnetic method. Dig Dis Sci 1992;37:1537-43.

[19]   Miranda JR, Baffa O, de Oliveira RB, Matsuda NM. An AC biosusceptometer to study gastric emptying. Med Phys 1992;19:445-8.

[20]   Andra W, Eitner K, Hergt R, Zeisberger M. Magnetic measurement of local transit time in the small intestine. Biomed Tech (Berl) 1997;42 Suppl:245-6.

[21]   Andra W, Danan H, Kirmsse W, Kramer HH, Saupe P, Schmieg R et al. A novel method for real-time magnetic marker monitoring in the gastrointestinal tract. Phys Med Biol 2000;45:3081-93.

[22]   Weitschies W, Wedemeyer J, Stehr R, Trahms L. Magnetic markers as a noninvasive tool to monitor gastrointestinal transit. IEEE Trans Biomed Eng 1994;41:192-5.

[23]   Weitschies W, Kotitz R, Cordini D, Trahms L. High-resolution monitoring of the gastrointestinal transit of a magnetically marked capsule. J Pharm Sci 1997;86:1218-22.

[24]   Weitschies W, Karaus M, Cordini D, Trahms L, Breitkreutz J, Semmler W. Magnetic marker monitoring of disintegrating capsules. Eur J Pharm Sci 2001;13:411-6.

[25]   Goodman K, Hodges LA, Band J, Stevens HN, Weitschies W, Wilson CG. Assessing gastrointestinal motility and disintegration profiles of magnetic tablets by a novel magnetic imaging device and gamma scintigraphy. Eur J Pharm Biopharm 2010;74:84-92.

[26]   Schlageter V, Besse PA, Popovic RS, Kucera P. Trackin system with five degrees of freedom using a 2D-array of Hall sensors and a permanent magnet. Sensors and Actuaters A: Physical 2001;92:37-42.

[27]   Schlageter V, Drljaca P, Popovic RS, Kucera P. A Magnetic Tracking System based on Highly Sensitive Integrated Hall Sensors. JSME International Journal Series C Mechanical Systems, Machine Elements and Manufacturing 2002;45:967-73.

[28]   Guignet R, Bergonzelli G, Schlageter V, Turini M, Kucera P. Magnet Tracking: a new tool for in vivo studies of the rat gastrointestinal motility. Neurogastroenterol Motil 2006;18:472-8.

[29]   Hiroz P, Schlageter V, Givel JC, Kucera P. Colonic movements in healthy subjects as monitored by a Magnet Tracking System. Neurogastroenterol Motil 2009;21:838-e57.

[30] Stathopoulos E, Schlageter V, Meyrat B, Ribaupierre Y, Kucera P. Magnetic pill tracking: a novel non-invasive tool for investigation of human digestive motility. Neurogastroenterol Motil 2005;17:148-54.

[31] Fynne L, Worsoe J, Gregersen T, Schlageter V, Laurberg S, Krogh K. Gastrointestinal transit in patients with systemic sclerosis. Scand J Gastroenterol 2011;46:1187-93.

[32] Fynne L, Worsoe J, Gregersen T, Schlageter V, Laurberg S, Krogh K. Gastric and small intestinal dysfunction in spinal cord injury patients. Acta Neurol Scand 2011.

[33] Gregersen T, Gronbaek H, Worsoe J, Schlageter V, Laurberg S, Krogh K. Effects of Sandostatin LAR on gastrointestinal motility in patients with neuroendocrine tumors. Scand J Gastroenterol 2011;46:895-902.

[34] Iddan G, Meron G, Glukhovsky A, Swain P. Wireless capsule endoscopy. Nature 2000;405:417.

[35] Cassilly D, Kantor S, Knight LC, Maurer AH, Fisher RS, Semler J et al. Gastric emptying of a non-digestible solid: assessment with simultaneous SmartPill pH and pressure capsule, antroduodenal manometry, gastric emptying scintigraphy. Neurogastroenterol Motil 2008;20:311-9.

[36] Baek NK, Sung IH, Kim DE. Frictional resistance characteristics of a capsule inside the intestine for microendoscope design. Proc Inst Mech Eng H 2004;218:193-201.

[37] Husebye E. The patterns of small bowel motility: physiology and implications in organic disease and functional disorders. Neurogastroenterol Motil 1999;11:141-6.

[38] Maqbool S, Parkman HP, Friedenberg FK. Wireless capsule motility: comparison of the SmartPill GI monitoring system with scintigraphy for measuring whole gut transit. Dig Dis Sci 2009;54:2167-74.

[39] Coupe AJ, Davis SS, Wilding IR. Variation in gastrointestinal transit of pharmaceutical dosage forms in healthy subjects. Pharm Res 1991;8:360-4.

Chapter 6.

Sacral nerve stimulation for faecal incontinence does not affect gastric contraction and small intestinal motility

J. Worsøe[*#], *J. Fassov*[□], *V. Schlageter*[+], *N.J.M. Rijkhoff*[#], *S. Laurberg*[*], *K. Krogh*[□]

[*]*Department of Surgery P, Aarhus University Hospital, Aarhus, Denmark*
[#]*Center for Sensory-Motor Interaction (SMI), Department of Health Science and Technology, Aalborg University, Denmark*
[□]*Neurogastroenterology Unit, Department of Hepatology and Gastroenterology, Aarhus University Hospital, Aarhus, Denmark*
[+]*Motilis Medica SA, Lausanne, Switzerland*

6.1 INTRODUCTION

Sacral nerve stimulation (SNS) was introduced for the treatment of idiopathic FI in 1995, and subsequently indications have spread to include FI of other etiologies. However, the mode of action is remains obscure [1,2]. Recent studies suggest that the effects of SNS are not mediated solely by the sacral pathways but also supraconal projections could be involved. Thus, reduced antegrade transport from the right colon and increased retrograde transport from the left colon at defecation has been demonstrated during SNS [3]. SNS also appears to inhibit the abnormal increase of postprandial rectal tone observed in some patients with idiopathic FI [4]. In a pilot study, SNS alleviated patients suffering from diarrhea predominant irritable bowel syndrome, a condition generally recognized to affect both the colon and the small intestine [5]. Even though the effects of SNS on the distal bowel have been extensively investigated, the effects on small intestinal transport has only been the subject of a single study, which did not show any change in small intestinal motility during SNS [6].

The recently developed magnet based Motility Tracking System (MTS-1) allows minor invasive, non irradiant assessment of gastric and small intestinal motility [7]. The aim of the present pilot study was to investigate motility patterns in the stomach and the small intestine during SNS in a group of patients successfully treated because of FI.

6.2 METHODS

6.2.1   Patients and efficacy of SNS

Among 178 FI patients having a permanent electrode and stimulator at our unit, 11 (all female, median (range) age 58 years (range: 41-74)) were recruited for the present study. SNS had reduced FI in all. The etiology of FI was partial sphincter lesion (n=3), incomplete spinal cord injury (n=3), sequelae from irradiation of some pelvic malignancy (n=1), and idiopathic FI (n=4). The median Wexner fecal incontinence score had decreased from 13 (range 11-20) before the PNE test, to 5.5 (range 2-13) one year after implantation and was 7 (range 3-9) at the date of inclusion. Median time since implantation was 42 months (range 25-100). At the time of the investigation, none of the patients took any medication affecting bowel function. Anal physiologic tests (including anal resting pressure, anal squeeze pressure, first sensation, desire to defecate, and maximal tolerable volume) were within normal range before and one year after implantation [8]. All participants signed informed written consent and the study was approved by the local scientific ethical committee Region Midt (M-20070273).

6.2.2   Magnet Tracking System, MTS-1

The Magnet Tracking System (MTS-1; Motilis, Lausanne, Switzerland) has been developed for use in a standard laboratory setting [9,10]. Subjects ingested a small magnetic pill (dimensions: 6x15 mm, weight: 1.62 g), which was tracked by a matrix of 4x4 magnetic field sensors separated by 5 cm and placed over the abdomen. The position of the sensor matrix with respect to anatomical landmarks was noted (the iliac spines, the intercostal angle, and the pubic bone) (Fig. 1A). With a sampling rate of 10 Hz, each sensor measured the magnetic induction dependent on the distance between the sensors and the magnetic pill and the orientation of the pill. The position and orientation of the magnetic pill was defined by 5 coordinates (position: x, y, z, and angle: θ, φ).The change in position coordinates reflected propagation of the magnet. The change of the angles reflected change in orientation, which correlated with the contraction frequency of the pertinent gastrointestinal segment. Data were analyzed on a computer running custom made software (MTS_Record, Motilis, Lausanne, Switzerland), which showed real time position and orientation of the magnetic pill (Fig. 1A, 1B). Before experiments, the sensors were calibrated by offsetting the earth's magnetic field. Artifacts due to respiration and movement were measured using accelerometers placed on the abdomen and the neck.

6.2.3   Protocol

Each patient was investigated twice. One week prior to the first investigation, the patients were randomized to the stimulator switched on or off. One week before the second investigation the stimulator was turned on or off opposite to the first investigation. The stimulator setting was carried out by a dedicated nurse in the department or the patient's general practitioner when convenient. The primary investigator and the patients were blinded to the setting of the stimulator. The patients kept a one-week bowel habit diary before the investigations. There was at least one month between the investigations during which the stimulator was set at the patient's usual level of chronic stimulation. Patients maintained their usual diet and daily activities during the study.

Following three hours fast, patients took the magnetic pill at 9 AM and recordings were performed for six hours. During investigations, subjects were placed in a bed with elevated head (>45°). They were encouraged not to talk or move but they could read or watch TV. The measurements were interrupted for small breaks on request.

6.2.4   Analysis and statistics

Motility patterns during and without SNS were analyzed with Motilis dedicated software for the upper gastro-intestinal tract (MTS_Tool, Motilis, Lausanne, Switzerland). The mean small intestinal propagation velocity for the two hour period following pyloric passage was computed. The pyloric passage was identified as the cessation of the 3 contractions min-1 pattern, typical for the stomach, the appearance of the duodenal arch, and the 8-11 contractions min-1 of the small intestine (Fig. 1A, 1B). The mean contraction frequencies of the stomach and the small intestine were calculated using the contractions with the highest amplitudes obtained when the magnet was performing very slow movements (i.e. when there was no progression of the magnet). The mean contraction frequency in the small intestine was calculated using only contractions with a frequency higher than 6 min-1. The frequency peaks were selected using a convolution of the Fast Fourier Transform with the "shape of a peak" described by a Gaussian function. The frequencies obtained during progression of the magnet were discarded, while frequencies obtained when the magnet did not progress were included. With this approach, the Doppler-effect (contraction frequency as a function of velocity of the magnet) was evaded.   A linear regression was used to derive the change in instantaneous contraction frequency during the first two hours after pyloric passage. An initial analysis of movement velocity of the magnetic pill had identified three types of movements, fast movements (> 15 cm min-1), slow movements (between 1.5 and 15 cm min-1), and very slow movement (<1.5 cm min-1). Based on this, an algorithm was developed for automatic classification of movements in the small intestine for the two hours after pyloric passage [11].

Numerical data are given as a median with total range. Statistical significance was tested with Wilcoxon's test (non-parametric test for paired data), and the level of significance was set at 0.05.

6.3 RESULTS

The investigation was well tolerated by all patients, and no adverse effects were encountered. Eight subjects were used for analysis of small intestinal motility. In three patients (one with partial SCI and two with idiopathic FI) there were insufficient data from the small intestine due to prolonged (more than 3 hours) retention of the magnet pill in the stomach (two investigations with SNS and one investigation with noSNS).

6.3.1   Gastric contraction frequency and gastric emptying

The mean gastric contraction frequency was 3.05 (range: 2.83-3.40 min-1) during SNS and 3.04 (range: 2.79-3.76 min-1) without SNS (ns). Median time for gastric emptying was 10 min (range 6-57) during SNS and 19 min (range 6-181) without SNS (ns).

6.3.2   Propagation velocities in the small intestine during SNS and without SNS

The two-hour motility data during SNS and postprandially are shown in table 6.1. The median two-hour propagation velocity was 1.6 cm min-1 (range: 1.2-2.8min-1) during SNS, and 1.7 cm min-1 (range: 0.8-3.7 min-1) without SNS (ns). During SNS, most small intestinal transit occurred with very fast movements (> 15 cm min-1) accounting for median 47% (range: 41-60%) of the distance in median 2.4% (range: 1.7-5.0%) of the time. Without SNS, 52 % (range: 18-73%) of the distance was covered with fast movements in median 3.3% (range: 0.8-3.3%) of the time.

6.3.3   Small intestinal contraction frequency

In the fasting state, the mean contraction frequency during the first two hours in the small intestine was 10.005 (range: 9.68-10.70) during SNS and 10.09 (range: 9.79-10.29) without SNS (ns). The mean contraction frequency decreased during the first two hours after pyloric passage both during SNS and without SNS. The decrease in contraction frequency was -0.965 min-1 per two hours (range: -0.22--1.75) during SNS and -0.845 min-1 per two hours (range: -0.58-0.28) without SNS(ns) (Table 6.2).

	SNS							
Subject ID	Fast movements (>15 cm min⁻¹) [cm]	[min]	Slow movements (<15 cm min⁻¹) [cm]	[min]	Very slow movements (<1.5 cm min⁻¹) [cm]	[min]	Progression [cm]	Velocity [cm min⁻¹]
1	83	3	85	31	29	87	197	1.6
2	149	5	110	47	15	69	274	2.3
3	115	3	41	18	37	101	193	1.6
4	75	3	67	34	39	85	181	1.5
5	85	2	53	25	41	93	179	1.5
6	155	6	174	57	10	58	339	2.8
7	66	2	50	18	25	99	141	1.2
8	115	5	88	39	44	76	247	2.1
Median	100	3	76	32.5	33	86	195	1.6
	No SNS							
	Fast movements (>15 cm min⁻¹) [cm]	[min]	Slow movements (<15 cm min⁻¹) [cm]	[min]	Very slow movements (<1.5 cm min⁻¹) [cm]	[min]	Progression [cm]	Velocity [cm min⁻¹]
1	59	3	95	38	11	71	165	1.4
2	111	4	111	45	25	71	247	2.1
3	18	1	40	23	41	98	99	0.8
4	321	10	86	38	35	74	442	3.7
5	77	3	53	28	28	88	158	1.3
6	240	8	161	50	28	59	429	3.6
7	107	4	33	12	25	104	165	1.4
8	150	6	45	17	39	99	234	2.0
Median	109	4	69.5	33	28	81	199.5	1.7

Table 6.1: Motility for two hours after pyloric passage during SNS and without SNS. Progression (cm) and duration (min) of fast (>15 cm min-1), slow (between 1.5 and 15 cm min-1), and very slow (<1.5 cm min-1) movements during fast and after a standard meal. The mean progression velocities during two hours are also given.

6.3.4   Bowel habit diary

The median number of bowel movements was 15 (range: 4-23) during SNS and 14 (range: 4-50) without SNS (ns). The median number of urge episodes was 2 (0-11) during SNS and 3 (0-21) without SNS. The median number of incontinence episodes was 0 (range: 0-6) during SNS and 0.5 (range: 0-21) without SNS (ns). The median number of soiling episodes was 1 (range: 0-7) during SNS and 1.5 (range: 0-7) without SNS (ns).

Subject ID	SNS		Without SNS	
	Frequency $[min^{-1}]$	2 hour delta $[min^{-1}\ cm^{-1}]$	Frequency $[min^{-1}]$	2 hour delta $[min^{-1}\ cm^{-1}]$
1	9.89	-0.22	10.07	-1.20
2	10.22	-0.51	10.14	-0.78
3	9.93	-0.59	9.95	-1.55
4	9.98	-0.58	10.29	0.28
5	10.03	-1.47	9.79	-0.58
6	10.70	-1.35	9.91	-0.91
7	10.53	-1.75	10.28	-1.00
8	9.68	-1.34	10.11	-0.62
median	10.005	-0.965	10.09	-0.845

Table 6.2:The two-hour mean contraction frequency in the small intestine, and the change rate in contraction frequency per two hours (delta) during SNS and without SNS. All recordings were made in fasting state.

6.4 DISCUSSION

This study utilized the Motility Tracking System (MTS-1) to investigate gastric and small intestinal motility in patients implanted with a sacral nerve stimulator because of FI of various origins. Gastric and small intestinal motility patterns during fast were recorded with the stimulator switched on or off after it had been turned off one week before the investigation. Data did not indicate any acute effect on gastric and small intestinal motility when turning off SNS. The propagation velocity, the distribution of periods with fast, slow movements, and very slow movements and the contraction frequencies were all unaffected by turning off SNS.

Colorectal physiology and anal physiology during SNS have been the subject of several studies. Data on anal squeeze pressure suggest some improvement of external anal sphincter function, while data on anal resting pressure and rectal sensation have been inconsistent [12-17]. A number of studies have shown that the effects of SNS are not restricted to the distal colorectum or the anal canal. Thus, bowel segments not receiving innervation from the sacral roots are affected by

SNS. Changes in postprandial rectal motility during SNS, have been demonstrated with 24-hour manometry and rectal impedance planimetry [4,18]. Furthermore, SNS appears to alleviate symptoms not only in idiopathic constipation but also in diarrhea predominant irritable bowel syndrome, the latter in which small intestinal transit is usually accelerated [5,19.20]. During SNS, an increased number of pancolonic pressure waves has been shown in constipated patients, and in patients treated for FI changes in colorectal transport in the right colon and increased colonic transit time have also been demonstrated [3,21]. Additionally, SNS alter cerebral evoked potential latency and reversibly reduce corticoanal excitability during transcranial magnetic stimulation indicating that dynamic brain changes could be involved [22-24]. Our group has found an increased activity in afferent vagal projections suggesting modulation of cerebral activity with stimulation [25]. Since SNS affects the proximal colon, alleviates symptoms in patients with IBS and causes modulation of cortical function, we hypothesized that gastric and small intestinal motility would change during SNS. Recently, Damgaard et al. used scintigraphy to investigate upper gastrointestinal function during stimulation and without stimulation, but no changes in gastric emptying or small intestinal transit were found [6].

It is suggested that SNS works by modulation of neural circuits rather than continuous nerve stimulation which can be switched off acutely. This could be the reason that no effect was seen after having the stimulator turned off for only one week, and also explain why there was no effect on symptoms recorded in the diary. Damgaard et al. also used a short period (1 week) to separate the change of stimulator setting which might explain their parallel results [6]. It contrasts previous studies that indicated rapid physiological and clinical effects, when the stimulator setting is changed. Thus, Kenefick et al. demonstrated and acute and reversible effect on rectal blood flow when turning off chronic stimulation [26]. Also, Emmanuel et al. have shown that stepwise increase of stimulation current acutely alters rectal mucosal blood flow [27]. Leroi et al. demonstrated significant fewer FI symptoms in the patients, when comparing a one month periods with the stimulator turned on versus a one month period with the stimulator turned off, and Michelsen et al. demonstrated an increase in Wexner and St. Mark's incontinence severity scores during the nightly switch off period, when comparing three week periods with the stimulator turned on versus off at night only [28,29].

Scintigraphy is the gold standard for evaluating gastric emptying and small intestinal motility. The method, however, does not evaluate contraction patterns. Manometry allows description of contraction patterns, but it is invasive and usually restricted to the duodenum and upper jejunum. MTS-1 is a novel alternative to the existing methods. It has been used in studies of fasting and postprandial motility in healthy subjects, and in subjects with gastrointestinal motility disorders [7,30,31]. The precision of MTS-1 depends on the position and orientation of the magnet with respect to the sensor matrix and the number of available sensors [10]. The present system can track the magnetic pill at distances of more than 200 mm and the absolute accuracy is approximately 1-2 cm, which is sufficient for anatomical localization. The amplitude of small back and forth movements can be measured more accurately (1-2 mm, rotation of 0.5°). Recently the validity of pyloric and ileocecal passage assessed with MTS-1 have been validated [7].

For practical and ethical reasons patients were not examined for more than 6 hours even if the magnet did not reach the caecum. Most often, a non-digestible object leaves the stomach during fast with an antral phase III MMC [32,33]. Accordingly, the time with recordings from the small intestine varied. Since motility patterns may change as the object moves distally in the small intestine standardization of position and initial progression with an antral phase III MMC were ensured by analyzing fasting small intestinal motility for two hours beginning just after pyloric passage [34,35].

There are several limitations to the present study. The small and heterogeneous group of patients studied carries a risk of type II error even in a cross-over design. Furthermore, the lack of significant difference in the symptoms when comparing periods during and without SNS, could be due to insufficient sensitivity of a bowel habit diary recorded for only one week. To avoid the risk of long term carry-over effects from SNS, future studies should compare motility before and after implantation of the stimulator.

6.5 CONCLUSION

In conclusion, turning off SNS for one week in patients treated for FI does not affect gastric or small intestinal motility.

REFERENCES

[1]   Amend B, Matzel KE, Abrams P, de Groat WC, Sievert KD. How does neuromodulation work. Neurourol Urodyn 2011;30:762-5.

[2]   Gourcerol G, Vitton V, Leroi AM, Michot F, Abysique A, Bouvier M. How sacral nerve stimulation works in patients with faecal incontinence. Colorectal Dis 2011;13:e203-11.

[3]   Michelsen HB, Christensen P, Krogh K, Rosenkilde M, Buntzen S, Theil J et al. Sacral nerve stimulation for faecal incontinence alters colorectal transport. Br J Surg 2008;95:779-84.

[4]   Michelsen HB, Worsoe J, Krogh K, Lundby L, Christensen P, Buntzen S et al. Rectal motility after sacral nerve stimulation for faecal incontinence. Neurogastroenterol Motil 2009.

[5]   Lundby L, Krogh K, Buntzen S, Laurberg S. Temporary sacral nerve stimulation for treatment of irritable bowel syndrome: a pilot study. Dis Colon Rectum 2008;51:1074-8.

[6] Damgaard M, Thomsen FG, Sorensen M, Fuglsang S, Madsen JL. The influence of sacral nerve stimulation on gastrointestinal motor function in patients with fecal incontinence. Neurogastroenterol Motil 2011;23:556-e207.

[7] Worsoe J, Fynne L, Gregersen T, Schlageter V, Christensen LA, Dahlerup JF et al. Gastric transit and small intestinal transit time and motility assessed by a magnet tracking system. BMC Gastroenterol 2011;11:145.

[8] Ryhammer AM, Laurberg S, Hermann AP. Test-retest repeatability of anorectal physiology tests in healthy volunteers. Dis Colon Rectum 1997;40:287-92.

[9] Schlageter V, Besse PA, Popovic RS, Kucera P. Trackin system with five degrees of freedom using a 2D-array of Hall sensors and a permanent magnet. Sensors and Actuaters A: Physical 2001;92:37-42.

[10] Schlageter V, Drljaca P, Popovic RS, Kucera P. A Magnetic Tracking System based on Highly Sensitive Integrated Hall Sensors. JSME International Journal Series C Mechanical Systems, Machine Elements and Manufacturing 2002;45:967-73.

[11] Hiroz P, Schlageter V, Givel JC, Kucera P. Colonic movements in healthy subjects as monitored by a Magnet Tracking System. Neurogastroenterol Motil 2009;21:838-e57.

[12] Michelsen HB, Buntzen S, Krogh K, Laurberg S. Rectal volume tolerability and anal pressures in patients with fecal incontinence treated with sacral nerve stimulation. Dis Colon Rectum 2006;49:1039-44.

[13] Vaizey CJ, Kamm MA, Turner IC, Nicholls RJ, Woloszko J. Effects of short term sacral nerve stimulation on anal and rectal function in patients with anal incontinence. Gut 1999;44:407-12.

[14] Tjandra JJ, Chan MK, Yeh CH, Murray-Green C. Sacral nerve stimulation is more effective than optimal medical therapy for severe fecal incontinence: a randomized, controlled study. Dis Colon Rectum 2008;51:494-502.

[15] Ganio E, Ratto C, Masin A, Luc AR, Doglietto GB, Dodi G et al. Neuromodulation for fecal incontinence: outcome in 16 patients with definitive implant. The initial Italian Sacral Neurostimulation Group (GINS) experience. Dis Colon Rectum 2001;44:965-70.

[16] Rosen HR, Urbarz C, Holzer B, Novi G, Schiessel R. Sacral nerve stimulation as a treatment for fecal incontinence. Gastroenterology 2001;121:536-41.

[17] Matzel KE, Bittorf B, Stadelmaier U, Hohenberger W. Sacral nerve stimulation in the treatment of faecal incontinence. Chirurg 2003;74:26-32.

[18] Altomare DF, Rinaldi M, Petrolino M, Monitillo V, Sallustio P, Veglia A et al. Permanent sacral nerve modulation for fecal incontinence and associated urinary disturbances. Int J Colorectal Dis 2004;19:203-9.

[19] Kamm MA, Dudding TC, Melenhorst J, Jarrett M, Wang Z, Buntzen S et al. Sacral nerve stimulation for intractable constipation. Gut 2010;59:333-40.

[20] Lu CL, Chen CY, Chang FY, Lee SD. Characteristics of small bowel motility in patients with irritable bowel syndrome and normal humans: an Oriental study. Clin Sci (Lond) 1998;95:165-9.

[21] Dinning PG, Fuentealba SE, Kennedy ML, Lubowski DZ, Cook IJ. Sacral nerve stimulation induces pan-colonic propagating pressure waves and increases defecation frequency in patients with slow-transit constipation. Colorectal Dis 2007;9:123-32.

[22] Sheldon R, Kiff ES, Clarke A, Harris ML, Hamdy S. Sacral nerve stimulation reduces corticoanal excitability in patients with faecal incontinence. Br J Surg 2005;92:1423-31.

[23] Harris ML, Singh S, Rothwell J, Thompson DG, Hamdy S. Rapid rate magnetic stimulation of human sacral nerve roots alters excitability within the cortico-anal pathway. Neurogastroenterol Motil 2008;20:1132-9.

[24] Giani I, Novelli E, Martina S, Clerico G, Luc AR, Trompetto M et al. The effect of sacral nerve modulation on cerebral evoked potential latency in fecal incontinence and constipation. Ann Surg 2011;254:90-6.

[25] Lundby L, Moller A, Buntzen S, Krogh K, Vang K, Gjedde A et al. Relief of fecal incontinence by sacral nerve stimulation linked to focal brain activation. Dis Colon Rectum 2011;54:318-23.

[26] Kenefick NJ, Emmanuel A, Nicholls RJ, Kamm MA. Effect of sacral nerve stimulation on autonomic nerve function. Br J Surg 2003;90:1256-60.

[27] Emmanuel AV, Kamm MA. Laser Doppler measurement of rectal mucosal blood flow. Gut 1999;45:64-9.

[28] Leroi A-, Parc Y, Lehur P-, Mion F, Barth X, Rullier E et al. Efficacy of sacral nerve stimulation for fecal incontinence: Results of a multicenter double-blind crossover study. Ann Surg 2005;242:662-9.

[29] Michelsen HB, Krogh K, Buntzen S, Laurberg S. A prospective, randomized study: switch off the sacral nerve stimulator during the night? Dis Colon Rectum 2008;51:538-40.

[30] Fynne L, Worsoe J, Gregersen T, Schlageter V, Laurberg S, Krogh K. Gastric and small intestinal dysfunction in spinal cord injury patients. Acta Neurol Scand 2011.

[31] Fynne L, Worsoe J, Gregersen T, Schlageter V, Laurberg S, Krogh K. Gastrointestinal transit in patients with systemic sclerosis. Scand J Gastroenterol 2011;46:1187-93.

[32] Mojaverian P, Chan K, Desai A, John V. Gastrointestinal transit of a solid indigestible capsule as measured by radiotelemetry and dual gamma scintigraphy. Pharm Res 1989;6:719-24.

[33] Cassilly D, Kantor S, Knight LC, Maurer AH, Fisher RS, Semler J et al. Gastric emptying of a non-digestible solid: assessment with simultaneous SmartPill pH and pressure capsule, antroduodenal manometry, gastric emptying scintigraphy. Neurogastroenterol Motil 2008;20:311-9.

[34] Kruis W, Azpiroz F, Phillips SF. Contractile patterns and transit of fluid in canine terminal ileum. Am J Physiol 1985;249:G264-70.

[35] Kellow JE, Borody TJ, Phillips SF, Tucker RL, Haddad AC. Human interdigestive motility: variations in patterns from esophagus to colon. Gastroenterology 1986;91:385-95.

Chapter   7.

Discussion

7.1 DISCUSSION

This thesis investigated the effect of DGN stimulation on FI symptoms and rectal motility. In contrast to SNS, electrical stimulation of the dorsal genital nerve allows stimulation of somatic afferents to the sacral spinal cord without implantation of an electrode. DGN stimulation could be a new treatment modality for FI, but it also allows new insight into the mechanisms of electrical modulation of bowel dysfunction.

The stimulation parameters for these experiments were derived from studies of DGN stimulation to achieve inhibition of the bladder [1,2]. It was anticipated that stimulation amplitudes of two times the genital-anal reflex threshold would be needed. This was applicable to SCI patients, but not to patients with idiopathic FI, as none of the patients could tolerate stimulation amplitudes that high. Both with therapeutic and acute stimulation, the stimulation amplitudes were set as high as the patient could tolerate.

The effects of DGN home stimulation was investigated in female patients with idiopathic FI. A significant decrease in the number of FI episodes was demonstrated as reflected in the Wexner and the St. Marks FI scores. Furthermore, the number of bowel movements and the percentage of bowel movements with urge also decreased significantly. This was, however, not associated with a significant improvement in quality of life. The effect of DGN stimulation appears to be maintained for weeks after stimulation is stopped. This is different from SNS where there seems to be an immediate on/ off effect [3-6]. This long term potentiation may indicate structural changes, i.e. either an up-regulation of receptors in the synapses or formation of new synapses. Finazzi-Agrò et al. recently demonstrated that PTNS, contrary to sham stimulation, modulates long latency somatosensory evoked potentials in patients with overactive bladder, suggesting involvement of central modulation [7]. DGN stimulation with low frequencies (1-2 Hz), as described by Binnie and Frizelle has another mechanism of action. Low frequency stimulation utilizes the genital-anal reflex to activate the sphincter muscles, which are then trained by repetitive contractions.

These findings have several implications. Further studies are needed to optimize the treatment and determine if the observed effects are strong enough to place DGN stimulation among other treatment modalities for FI. Being non-invasive and without risk, DNG stimulation may be offered on a liberal basis to patients with idiopathic FI and, perhaps, also patients with FI secondary to disease or patients with irritable bowel syndrome. However, this should be preceded by well-conducted randomized trials.

DGN home stimulation had no effect on anal resting pressure, anal squeeze pressure, or rectal volume tolerability. Sphincter manometry and rectal volume tolerability were within reference values for women in western Denmark both before and during stimulation [8]. Perhaps three weeks is insufficient time to obtain any significant improvement of sphincter function. While low frequency DGN stimulation improves sphincter function, studies of SNS for the treatment of FI are inconsistent with respect to the effect on sphincter function and rectal volume tolerability [6,9-16]. This suggests that it is not essential to improve sphincter function to reduce FI symptoms in patients with idiopathic FI using DGN stimulation at 20 Hz.

IP was used to study the rectal CSA in idiopathic FI patients and patients with a supraconal SCI. The acute effect of DGN stimulation was studied using a probe which could measure one CSA at different distension pressures. DGN stimulation did not change rectal CSA in idiopathic FI patients. It is possible that the lack of response during acute stimulation was caused by too low stimulation amplitude. Nevertheless, most of the patients, including some with very low stimulation amplitudes just above the reflex threshold, experienced that FI symptoms decreased during stimulation. This indicates that amplitudes below two times the reflex threshold are sufficient to treat FI. If the stimulation amplitude can be lower, stimulation will be less bothersome. This will likely increase patient compliance, as more patients will tolerate the treatment. Furthermore, a placebo effect during DGN stimulation is likely as previously described in studies of transcutaneous stimulation for reduction of pain [17,18].

The CSAs measured without DGN stimulation in idiopathic FI patients were similar to previous findings in healthy subjects obtained with similar distension protocols. Thus, the range of rectal CSA was from 7 to 41 cm2 corresponding to findings [19,20]. Even though no difference in rectal pressure-CSA relation between healthy controls and patients with idiopathic FI could be found, it is possible that an increase of the pressure-CSA relation within the individual patient could affect continence mechanism and reduce FI symptoms.

Stimulation of the genital-anal reflex, in patients without suprasacral inhibition, resulted in a decrease in rectal CSA and rectal compliance expressed as the pressure-CSA relation. These findings were in contrast to those described with acute DGN stimulation of neurogenic bladder dysfunction where stimulation was able to inhibit bladder contractions [2]. Experiences from the bladder may not be directly applied to the gastrointestinal tract, which is also controlled by other factors including the enteric nervous system and humeral agents. However, these results are also in contrast to results by Chung et al. (presented only in a review by Craggs et al. [21]) who found that DGN stimulation elicited an inhibitory effect on rectal tone in patients with supraconal SCI. How this discrepancy should be explained is not clear. With ambiguous findings, the investigation should be repeated by others to confirm the results.

SCI patients included in this study had larger rectal CSAs, measured without DGN stimulation, compared to existing rectal CSA data in SCI patients [20]. This difference was probably due to difference in timing of the investigation – median 19 years after the injury vs. within the first year since injury.

In the second part of the thesis, MTS-1 was tested for determination of gastric emptying and small intestinal transit with simultaneous use of PillCam. The validity of gastric emptying and small intestinal transit was good and inter observer variations were small.

The precision of MTS-1 depends on the position and orientation of the magnet with respect to the sensor matrix. With only one sensor positioned 100 mm from the magnetic pill, the positioning error in the frontal plane is 10 mm, but changes in orientation of only 1-2 degrees can be detected [22]. This error is reduced by adding more sensors in a matrix, and the currently used present system can track the magnetic pill at distances of more than 200 mm with a positioning error of 10 mm.

Compared to scintigraphy, MTS-1 has the advantage of being without risk from exposure to radiation. This is especially important if children are investigated. Scintigraphy, however, allows determination of gastric emptying for both solids and liquids whereas the magnetic pill is only solid. Apart from giving the segmental transit times, capsule endoscopy has also been used for description of small intestinal motility patterns [23,24]. The size of the PillCam has been suspected to interfere with motility, but data from the present study contradict this, as transit times with the specially constructed magnet-PillCam unit did not differ from those obtained with the much smaller magnetic pill.

Currently, manometry is the best way to investigate antroduodenal and small intestinal motility patterns [25,26](merged)25. MTS-1 was expected to identify the phase III in the migrating motor complexes (MMC) and in the recordings during fast several examples of what must be MMC phase III were seen. However, future studies combining manometry and MTS are needed to validate changes in MMC seen by MTS. The propagated distance of the magnetic pill was the same during fast and postprandially. During fast the contraction frequency decreased in the aboral direction and this was even more pronounced postprandially. This probably reflects the small intestine adapting to intake of food. In accordance with this, the mean contraction frequency in the small intestine increased postprandially.

A wireless motility capsule (Smartpill, SmartPill Corporation, Buffalo, NY, USA) has recently been introduced. It is for ambulatory use, and measures pressure, pH and temperature throughout the gastrointestinal tract [27,28]. The Smartpill provides reliable information about gastric emptying, small intestinal transit, total colonic transit, and some contraction patterns [29,30]. In contrast to MTS-1, it does not provide data of segmental colonic transit times. The Smartpill is equally non-invasive and the ambulatory use is a major advantage compared to MTS-1.

Finally MTS-1 was used to investigate small intestinal motility in patients implanted with a stimulator because of idiopathic FI. With this method it was not possible to show any effect from SNS on small intestinal motility. The protocol was designed for investigation of motility during SNS compared to motility without stimulation. One week ON versus one week OFF periods were compared, since previous studies points to a rapid effect, when the stimulator setting is changed [3,14,31,32]. When assessing the bowel habit diaries in this study, there was no significant difference in incontinence symptoms, and possibly, a one week diary for monitoring a wash out effect is too short to record any difference in symptom score due to the relative low number of events.

It has been suggested that SNS may have an effect on intestinal motility and function in general. Primarily, it has been proposed that SNS alters motility in the right colon [33]. Common extrinsic innervation by the vagus nerve implies that SNS could also affect small intestinal motility. Furthermore, changes in postprandial rectal motility during SNS, have been demonstrated with 24-hours manometry and rectal impedance planimetry [34,35]. Others have found that SNS reversibly reduced corticoanal excitability during transcranial magnetic stimulation indicating that dynamic brain changes could be involved [36,37]. Secondly, SNS appear to alleviate patients suffering from both constipation and irritable bowel syndrome [38,39]. In patients with diarrhea-predominant IBS, small intestinal transit is accelerated [40]. These studies could indicate that SNS influences autonomous regulation subsequently affecting whole gut motility and sensory function. Even so, it could not be demonstrated with MTS-1.

7.2 CONCLUSIONS

Based on studies in the present thesis the following is concluded:
• DGN home stimulation reduces the number of FI episodes in most patients suffering from idiopathic FI.
• Reduction of FI episodes is obtained even at low stimulation amplitudes equal to the genital anal reflex threshold.
• DGN home stimulation does not affect sphincter function and rectal volume tolerability.
• DGN stimulation does not elicit an acute effect on rectal CSA, the rectal volume-pressure relation or the wall tension in patients suffering from idiopathic FI.
• Acute DGN stimulation reduces the rectal CSA in patients with supraconal SCI.
• MTS-1 can be used for determination of gastric emptying and small intestinal transit time.
• Using MTS-1 it is possible to distinguish between the mean contraction frequency of the small intestine in the fasting state and in the postprandial state.

• Data obtained using MTS-1 do not show any effect of SNS on proximal small intestinal motility.

7.3 PERSPECTIVES

Regarding DGN stimulation, the main task is to establish good evidence of a clinical effect. This may help to establish the role of DGN stimulation in the treatment of FI, to establish optimal stimulation parameters and stimulation protocols, and possibly to learn more about the mechanism of action. The initial experiences from chapter I indicate that DGN stimulation improves FI symptoms. A randomized double blinded study with sham stimulation is needed to clarify if DGN stimulation is better than placebo. Optimization of stimulation regime should also be investigated in a randomized design, with testing of different stimulation parameters and frequency of stimulation sessions. Short-term follow-up (3 weeks) indicates that the effect is maintained, but long-term follow-up should also be described. Furthermore, DGN stimulation should be assessed against other modes of electrical stimulation for management of FI, i.e. PTNS and the test stimulation period prior to SNS.

New techniques for description of anorectal function and colonic motility may yield insight to the mechanism of action of DGN stimulation and SNS. The functional lumen imaging probe (FLIP) can be used for investigation of the anal sphincters and might give more detailed information about tone and pressure gradients during DGN stimulation [41]. Scintigraphic studies have indicated an effect from SNS on colorectal transport. Novel high resolution manometry catheters will allow a highly detailed characterization of colorectal motility during stimulation [42].

In this thesis, MTS-1 was introduced as a new non-invasive system to yield information about transit time and motility patterns. Further development of the equipment including an ambulatory setup and improved software for analysis will enable whole gut investigations with more detailed data analysis. Results presented in study V are considered preliminary. Further evaluation of data is needed before drawing more firm conclusions about effects on SNS on upper gastrointestinal function.

REFERENCES

[1]   Previnaire JG, Soler JM, Perrigot M, Boileau G, Delahaye H, Schumacker P et al. Short-term effect of pudendal nerve electrical stimulation on detrusor hyperreflexia in spinal cord injury patients: importance of current strength. Paraplegia 1996;34:95-9.

[2]   Hansen J, Media S, Nohr M, Biering-Sorensen F, Sinkjaer T, Rijkhoff NJ. Treatment of neurogenic detrusor overactivity in spinal cord injured patients by conditional electrical stimulation. J Urol 2005;173:2035-9.

[3]   Michelsen HB, Krogh K, Buntzen S, Laurberg S. A prospective, randomized study: switch off the sacral nerve stimulator during the night? Dis Colon Rectum 2008;51:538-40.

[4]   Queralto M, Portier G, Cabarrot PH, Bonnaud G, Chotard JP, Nadrigny M et al. Preliminary results of peripheral transcutaneous neuromodulation in the treatment of idiopathic fecal incontinence. Int J Colorectal Dis 2006;21:670-2.

[5]   de la Portilla F, Rada R, Vega J, Gonzalez CA, Cisneros N, Maldonado VH. Evaluation of the use of posterior tibial nerve stimulation for the treatment of fecal incontinence: preliminary results of a prospective study. Dis Colon Rectum 2009;52:1427-33.

[6]   Frizelle FA, Gearry RB, Johnston M, Barclay ML, Dobbs BR, Wise C et al. Penile and clitoral stimulation for faecal incontinence: External application of a bipolar electrode for patients with faecal incontinence. Colorectal Disease 2004;6:54-7.

[7]   Finazzi-Agro E, Rocchi C, Pachatz C, Petta F, Spera E, Mori F et al. Percutaneous tibial nerve stimulation produces effects on brain activity: study on the modifications of the long latency somatosensory evoked potentials. Neurourol Urodyn 2009;28:320-4.

[8]   Ryhammer AM, Laurberg S, Hermann AP. Test-retest repeatability of anorectal physiology tests in healthy volunteers. Dis Colon Rectum 1997;40:287-92.

[9]   Matzel KE, Stadelmaier U, Hohenfellner M, Gall FP. Electrical stimulation of sacral spinal nerves for treatment of faecal incontinence. Lancet 1995;346:1124-7.

[10]   Vaizey CJ, Kamm MA, Turner IC, Nicholls RJ, Woloszko J. Effects of short term sacral nerve stimulation on anal and rectal function in patients with anal incontinence. Gut 1999;44:407-12.

[11]   Vaizey CJ, Kamm MA, Roy AJ, Nicholls RJ. Double-blind crossover study of sacral nerve stimulation for fecal incontinence. Dis Colon Rectum 2000;43:298-302.

[12]   Ganio E, Ratto C, Masin A, Luc AR, Doglietto GB, Dodi G et al. Neuromodulation for fecal incontinence: outcome in 16 patients with definitive implant. The initial Italian Sacral Neurostimulation Group (GINS) experience. Dis Colon Rectum 2001;44:965-70.

[13] Tjandra JJ, Chan MK, Yeh CH, Murray-Green C. Sacral nerve stimulation is
      more effective than optimal medical therapy for severe fecal incontinence: a
      randomized, controlled study. Dis Colon Rectum 2008;51:494-502.

[14] Leroi AM, Parc Y, Lehur PA, Mion F, Barth X, Rullier E et al. Efficacy of
      sacral nerve stimulation for fecal incontinence: Results of a multicenter
      double-blind crossover study. Ann Surg 2005;242:662-9.

[15] Michelsen HB, Buntzen S, Krogh K, Laurberg S. Rectal volume tolerability
      and anal pressures in patients with fecal incontinence treated with sacral
      nerve stimulation. Dis Colon Rectum 2006;49:1039-44.

[16] Binnie NR, Kawimbe BM, Papachryosostomou M, Smith AN. Use of the
      pudendo-anal reflex in the treatment of neurogenic faecal incontinence. Gut
      1990;31:1051-5.

[17] Conn IG, Marshall AH, Yadav SN, Daly JC, Jaffer M. Transcutaneous
      electrical nerve stimulation following appendicectomy: the placebo effect.
      Ann R Coll Surg Engl 1986;68:191-2.

[18] Brosseau L, Milne S, Robinson V, Marchand S, Shea B, Wells G et al.
      Efficacy of the transcutaneous electrical nerve stimulation for the treatment
      of chronic low back pain: a meta-analysis. Spine (Phila Pa 1976)
      2002;27:596-603.

[19] Dall FH, Jorgensen CS, Houe D, Gregersen H, Djurhuus JC. Biomechanical
      wall properties of the human rectum. A study with impedance planimetry.
      Gut 1993;34:1581-6.

[20] Krogh K, Mosdal C, Gregersen H, Laurberg S. Rectal wall properties in
      patients with acute and chronic spinal cord lesions. Dis Colon Rectum
      2002;45:641-9.

[21] Craggs MD, Balasubramaniam AV, Chung EA, Emmanuel AV. Aberrant
      reflexes and function of the pelvic organs following spinal cord injury in
      man. Auton Neurosci 2006;126-127:355-70.

[22] Sasaki Y, Hada R, Nakajima H, Munakata A. Difficulty in estimating
      localized bowel contraction by colonic manometry: a simultaneous recording
      of intraluminal pressure and luminal calibre. Neurogastroenterol Motil
      1996;8:247-53.

[23] Sorensen M, Rasmussen OC, Tetzschner T, Christiansen J. Physiological
      variation in rectal compliance Br J Surg 1992;79:1106-8.

[24] Oettle GJ, Heaton KW. "Rectal dissatisfaction" in the irritable bowel
      syndrome. A manometric and radiological study. Int J Colorectal Dis
      1986;1:183-5.

[25] Hansen MB. Small intestinal manometry. Physiol Res 2002;51:541-56.

[26] Husebye E. The patterns of small bowel motility: physiology and implications in organic disease and functional disorders. Neurogastroenterol Motil 1999;11:141-6.

[27] Huizinga JD. Electrophysiology of human colon motility in health and disease. Clin Gastroenterol 1986;15:879-901.

[28] Read MG, Read NW. Role of anorectal sensation in preserving continence. Gut 1982;23:345-7.

[29] Shefchyk SJ. Modulation of excitatory perineal reflexes and sacral striated sphincter motor neurons during micturition in the cat. In: Rudomin P, Romo R, Mendell LM, editors. Presynaptic inhibition and neural control., New York: Oxford University Press; 1998, p. 398-405.

[30] Sundin T, Carlsson CA, Kock NG. Detrusor inhibition induced from mechanical stimulation of the anal region and from electrical stimulation of pudendal nerve afferents. An experimental study in cats. Invest Urol 1974;11:374-8.

[31] Kenefick NJ, Emmanuel A, Nicholls RJ, Kamm MA. Effect of sacral nerve stimulation on autonomic nerve function. Br J Surg 2003;90:1256-60.

[32] Emmanuel AV, Kamm MA. Laser Doppler measurement of rectal mucosal blood flow. Gut 1999;45:64-9.

[33] Michelsen HB, Christensen P, Krogh K, Rosenkilde M, Buntzen S, Theil J et al. Sacral nerve stimulation for faecal incontinence alters colorectal transport. Br J Surg 2008;95:779-84.

[34] Michelsen HB, Maeda Y, Lundby L, Krogh K, Buntzen S, Laurberg S. Retention test in sacral nerve stimulation for fecal incontinence. Dis Colon Rectum 2009;52:1864-8.

[35] Altomare DF, Rinaldi M, Petrolino M, Monitillo V, Sallustio P, Veglia A et al. Permanent sacral nerve modulation for fecal incontinence and associated urinary disturbances. Int J Colorectal Dis 2004;19:203-9.

[36] Sheldon R, Kiff ES, Clarke A, Harris ML, Hamdy S. Sacral nerve stimulation reduces corticoanal excitability in patients with faecal incontinence. Br J Surg 2005;92:1423-31.

[37] Harris ML, Singh S, Rothwell J, Thompson DG, Hamdy S. Rapid rate magnetic stimulation of human sacral nerve roots alters excitability within the cortico-anal pathway. Neurogastroenterol Motil 2008;20:1132-9.

[38] Kamm MA, Dudding TC, Melenhorst J, Jarrett M, Wang Z, Buntzen S et al. Sacral nerve stimulation for intractable constipation. Gut 2010;59:333-40.

[39] Lundby L, Krogh K, Buntzen S, Laurberg S. Temporary sacral nerve stimulation for treatment of irritable bowel syndrome: a pilot study. Dis Colon Rectum 2008;51:1074-8.

[40] Lu CL, Chen CY, Chang FY, Lee SD. Characteristics of small bowel motility in patients with irritable bowel syndrome and normal humans: an Oriental study. Clin Sci (Lond) 1998;95:165-9.

[41] McMahon BP, Frokjaer JB, Liao D, Kunwald P, Drewes AM, Gregersen H. A new technique for evaluating sphincter function in visceral organs: application of the functional lumen imaging probe (FLIP) for the evaluation of the oesophago-gastric junction. Physiol Meas 2005;26:823-36.

[42] Arkwright JW, Underhill ID, Maunder SA, Blenman N, Szczesniak MM, Wiklendt L et al. Design of a high-sensor count fibre optic manometry catheter for in-vivo colonic diagnostics. Opt Express 2009;17 22423-31.

About the Author

The author received his MD from Aarhus University Hospital 2004. After working as inturn and in general practive for 1 year, he worked at the Colorectal Dept. at Aarhus Universtity Hospital for one year before starting the Phd-study at the department of sensorymotor interaction at Aalborg University. Currently undergoes specialty training in gastrointestinal surgery and continue to do research in electrical stimulation for the treatment of defecatory disorders.

www.ingramcontent.com/pod-product-compliance
Lightning Source LLC
Chambersburg PA
CBHW061322220326
41599CB00026B/4992